# Advancing Diversity, Equity, Inclusion, and Accessibility in the Leadership of Competed Space Missions

Committee on Increasing Diversity and Inclusion in the Leadership of Competed Space Missions

Space Studies Board
Division on Engineering and Physical Sciences

Board on Science Education
Division of Behavioral and Social Sciences and Education

A Consensus Study Report of

*The National Academies of*
SCIENCES · ENGINEERING · MEDICINE

THE NATIONAL ACADEMIES PRESS
*Washington, DC*
**www.nap.edu**

**THE NATIONAL ACADEMIES PRESS** **500 Fifth Street, NW** **Washington, DC 20001**

This activity was supported by Contract NNH17CB02B with the National Aeronautics and Space Administration. Any opinions, findings, conclusions, or recommendations expressed in this publication do not necessarily reflect the views of any organization or agency that provided support for the project.

Cover design by Timothy Warchocki. *Background:* Astronaut image ISS059-E-36323 of the Mississippi Delta; courtesy of the Earth Science and Remote Sensing Unit, NASA Johnson Space Center, https://eol.jsc.nasa.gov. *Top row, second from left:* X-ray image of the Sun by NASA's Nuclear Spectroscopic Telescope Array, or NuSTAR, overlaid on a picture taken by NASA's Solar Dynamics Observatory (SDO); courtesy of NASA/JPL-Caltech/GSFC. *Middle row, left:* Dr. Ali Guarneros-Luna holds CubeSat hardware; courtesy of NASA/Dominic Hart. *Middle row, fifth from left:* Image of the Orion Nebula taken by the Hubble Space Telescope in January 2006; courtesy of NASA/JPL-Caltech/GSFC. *Middle row, seventh from left:* March 2, 2015, astronaut image ISS042E340851; courtesy of NASA. *Middle row, eighth from left:* Alice Bowman, the Mission Operations Manager for NASA's New Horizons mission, at work in the Mission Operations Center on December 6, 2014; courtesy of SwRI/JHUAPL. *Bottom row, left:* Dr. Temilola Fatoyinbo presents some of the data visualizations and Earth Science implications of current satellite missions to Vice President Kamala Harris using Goddard's Hyperwall on November 5, 2021, at the NASA Goddard Space Flight Center in Greenbelt, Maryland; courtesy of NASA/Taylor Mickal. *Bottom row, second from left:* Nelson Huang, Manuel Soriano, and John Trinh stand in front of NASA's Psyche mission; courtesy of NASA/JPL-Caltech/M. Soriano. *Bottom row, third from left:* Image of Jupiter taken by the Hubble Space Telescope on June 27, 2019; courtesy of NASA, ESA, A. Simon (Goddard Space Flight Center), and M.H. Wong (University of California, Berkeley). *Bottom row, seventh from left:* Dr. Lindy Elkins-Tanton, principal investigator of NASA's Psyche mission from Arizona State University gives remarks during a briefing discussing small bodies missions, December 31, 2018; courtesy of NASA/Bill Ingalls.

International Standard Book Number-13: 978-0-309-27317-6
International Standard Book Number-10: 0-309-27317-X
Digital Object Identifier: https://doi.org/10.17226/26385
Library of Congress Control Number: 2022940422

Additional copies of this publication are available from the National Academies Press, 500 Fifth Street, NW, Keck 360, Washington, DC 20001; (800) 624-6242 or (202) 334-3313; http://www.nap.edu.



Printed in the United States of America

Suggested citation: National Academies of Sciences, Engineering, and Medicine. 2022. *Advancing Diversity, Equity, Inclusion, and Accessibility in the Leadership of Competed Missions.* Washington, DC: The National Academies Press. https://doi.org/10.17226/26385.

*The National Academies of*
SCIENCES • ENGINEERING • MEDICINE

**COMMITTEE ON INCREASING DIVERSITY AND INCLUSION IN THE LEADERSHIP OF COMPETED SPACE MISSIONS**

FRANCES BAGENAL, NAS,[1] University of Colorado Boulder, *Co-Chair*
WANDA E. WARD, University of Illinois at Urbana-Champaign, *Co-Chair*
CLAUDE R. CANIZARES, NAS, Massachusetts Institute of Technology
ERIN CECH, University of Michigan
ANNMARIE ELDERING, Jet Propulsion Laboratory
ELENA A. EROSHEVA, University of Washington
ORLANDO FIGUEROA, Orlando Leadership, LLC
ALLISON N. JAYNES, University of Iowa
LINDSEY MALCOM-PIQUEUX, California Institute of Technology
DAVID J. MCCOMAS, Princeton University
CHRISTOPHER B. NEWMAN, Azusa Pacific University
KUNIO M. SAYANAGI, Hampton University
KEIVAN G. STASSUN, Vanderbilt University
LOUIS-GREGORY STROLGER, Space Telescope Science Institute

***Staff***

SANDRA GRAHAM, Senior Program Officer, Space Studies Board, *Co-Study Director*
TIFFANY E. TAYLOR, Board on Science Education, *Co-Study Director*
ABIGAIL SHEFFER, Senior Program Officer, Space Studies Board
GAYBRIELLE HOLBERT, Senior Program Assistant, Space Studies Board

COLLEEN N. HARTMAN, Director, Space Studies Board
HEIDI SCHWEINGRUBER, Director, Board on Science Education

[1] Member, National Academy of Sciences.

## SPACE STUDIES BOARD



[1] Member, National Academy of Sciences.
[2] Member, National Academy of Engineering.

**BOARD ON SCIENCE EDUCATION**



[1] Member, National Academy of Sciences.

# Acknowledgments

This Consensus Study Report reflects the invaluable contributions of many individuals who provided their expertise, including those who served on the committee, the staff of the National Academies of Sciences, Engineering, and Medicine, and many other experts who participated in discussions with the committee.

This report was made possible by sponsorship from the National Aeronautics and Space Administration (NASA). In particular, the committee would like to first acknowledge both Thomas Zurbuchen, Associate Administrator of NASA's Science Mission Directorate, and Michael New, Deputy Associate Administrator of Research in NASA's Science Mission Directorate, for requesting the study, for being on hand to provide resources, answer committee questions, and to offer insight throughout the study process.

Over the course of the study, the committee held several public fact-finding meetings, of which committee members greatly benefited from. Many presentations and discussions were held with many individuals, including Thomas Wagner, NASA Science Mission Directorate; Deborah Amato, NASA Goddard Space Flight Center; Cathy Olkin, Southwest Research Institute; Sarah Lipscy, Ball Aerospace; Erika Hamden, University of Arizona; Nicola Fox, NASA Science Mission Directorate; Elizabeth Cole, University of Michigan; Susan White, American Institute of Physics; Christopher Keane, American Geosciences Institute; Louis Barbier, NASA Office of the Chief Scientist; Kelly Case, Jet Propulsion Laboratory; Sabrina Feldman, Jet Propulsion Laboratory; Bruce Jakosky, University of Colorado Boulder; Sean Jones, National Science Foundation; Stephen Meacham, National Science Foundation; Bernice Anderson, National Science Foundation; Marie Bernard, National Institutes of Health; Shamala Srinivas, National Cancer Institute; Caron Lyman, National Cancer Institute; Lauren Moen, National Heart, Lung, and Blood Institute; Alison Gammie, National Institute of General Medical Sciences; Stephen Fiore, University of Central Florida; Janet Vertesi, Princeton University; Stephanie Pfirman, Arizona State University; Jim Burch, Southwest Research Institute; Lisa Wood, Ball Aerospace; Lisa Callahan, Lockheed Martin; Mike Ryschkewitsch, Johns Hopkins Applied Physics Laboratory; Lindy Elkins-Tanton, Arizona State University; Tuija Pulkkinen, University of Michigan; Harlan Spence, University of New Hampshire; Louisa Koch, National Oceanic and Atmospheric Administration; Ping Ge, Department of Energy; Victor McCrary, National Science Board; Heidi Jensen, NASA Science Mission Directorate, Lorenzo Pappas, NASA Science Mission Directorate; Brenda Manuel, former NASA Associate Administrator for Diversity and Equal Opportunity; Elaine Ho, NASA Office of STEM Engagement; Kris Brown, NASA Office of STEM Engagement; Torry Johnson, NASA Office of STEM Engagement; Richard Gilmore, NASA Office of STEM Engagement; Philip Sakimoto; University of Notre Dame; Louie Lopez, Department of Defense; and Charles Bolden, former NASA Administrator.

The committee is also extremely grateful to Mithuna Srinivasan, Heather Sawyer, Ann Davoren, Elena Navarro, and Julie Kubelka from the National Opinion Research Center (NORC) at the University of Chicago who consulted with us to conduct a qualitative study on the competed space mission proposal process and the experiences of principal investigators for the committee's use in drafting the report.

Lastly, the committee wishes to extend its gratitude to the staff of both the Division on Engineering and Physical Sciences (DEPS) and the Division of Behavioral and Social Sciences and Education (DBASSE), in particular to Sandra Graham of the Space Studies Board (DEPS/SSB) and Tiffany Taylor of the Board on Science Education (DBASSE/BOSE), who as co-study directors, shepherded the committee throughout the entire process and played a key role in the report preparation and review process; to Abigail Sheffer (DEPS/SSB), who provided critical assistance in project direction and report review; to Gaybrielle Holbert (DEPS/SSB), whose expert administrative leadership and support enabled smooth meetings and report production; to Mia Brown (DEPS/SSB), who stepped in to provide timely help and support to prepare the report for review; and to Colleen N. Hartman (director of SSB) and Heidi Schweingruber (director of BOSE), whose expert advice, strategic thinking, and helpful suggestions throughout the entire study process ensured a timely and impactful report. Liz Panos of the DEPS staff expertly guided the report through the National Academies review process and Catherine Gruber of the DEPS staff provided invaluable editorial assistance.

# Acknowledgment of Reviewers

This Consensus Study Report was reviewed in draft form by individuals chosen for their diverse perspectives and technical expertise. The purpose of this independent review is to provide candid and critical comments that will assist the National Academies of Sciences, Engineering, and Medicine in making each published report as sound as possible and to ensure that it meets the institutional standards for quality, objectivity, evidence, and responsiveness to the study charge. The review comments and draft manuscript remain confidential to protect the integrity of the deliberative process.

We thank the following individuals for their review of this report:

Elizabeth R. Cole, University of Michigan,
Stephen M. Fiore, University of Central Florida,
Lennard A. Fisk, NAS,[1] University of Michigan,
Chelle L. Gentemann, Farallon Institute,
Martha S. Gilmore, Wesleyan University,
Erika T. Hamden, University of Arizona,
Donald G. James, NASA (retired),
Roy Maizel, NASA (retired),
Jeffrey Newmark, NASA Goddard Space Flight Center,
Becky Wai-ling Packard, Mount Holyoke College,
Suzanne Smrekar, Jet Propulsion Laboratory,
Michael Turner, NAS, University of Chicago, and
Lisa D. White, University of California, Berkeley.

[1] Member, National Academy of Sciences.

Although the reviewers listed above provided many constructive comments and suggestions, they were not asked to endorse the conclusions or recommendations of this report nor did they see the final draft before its release. The review of this report was overseen by Wesley L. Harris, NAE,[2] Massachusetts Institute of Technology, and Martha Haynes, NAS, Cornell University. They were responsible for making certain that an independent examination of this report was carried out in accordance with the standards of the National Academies and that all review comments were carefully considered. Responsibility for the final content rests entirely with the authoring committee and the National Academies.

[2] Member, National Academy of Engineering.

# Contents

# Summary

Several studies have demonstrated that multiple forms of diversity are beneficial to the creativity, innovation, and impact of science teams (Hong and Page 2004; Campbell et al. 2013; Freeman and Huang 2014a,b). Although social scientists continue to try to understand how specific forms of diversity (e.g., racial/ethnic, gender, career stage diversity) relate to various aspects of team performance, it is generally understood that, when engaged productively in an inclusive environment, diverse perspectives, experiences, and backgrounds can strengthen teams and lead to better science (Sommers 2006; Diaz-Garcia et al. 2013; Page 2008, 2017). Accordingly, the National Aeronautics and Space Administration's (NASA's) *Vision for Scientific Excellence* notes, "diversity is a key driver of innovation and more diverse organizations are more innovative.... NASA believes in the importance of diverse and inclusive teams to tackle strategic problems and maximize scientific return" (NASA 2020b).

At the request of NASA, the National Academies of Sciences, Engineering, and Medicine convened a study committee with the task of recommending actions to increase diversity, equity, inclusion, and accessibility (DEIA) in the leadership of space mission proposals submitted to competed space mission programs in NASA's Science Mission Directorate (SMD). The expertise of the committee spanned a wide range of experience from both scientists involved in NASA missions as well as the social sciences. The Committee on Increasing Diversity and Inclusion in the Leadership of Competed Space Missions approached the task through meetings with representatives of different groups and institutions and extensive examination of relevant literature, as well as by commissioning a study involving structured interviews to evaluate what barriers a sample of mission principal investigators (PIs) experienced in putting together a mission proposal.

The role of a PI on a competed mission is substantial. The job requires considerable experience in the scientific research field, knowledge of mission design, building of hardware and concepts of mission operation, plus the ability to lead and manage a large team of scientists, engineers, and administrators. Currently, PIs tend to be at senior levels of their career and thus, the demographics of mission PIs most likely reflect the population at such senior levels of their career. Consequently, the committee needed to examine the full pathway into Earth and space sciences careers to identify opportunities for diversifying the pool of potential proposal leadership teams, starting with decisions by young people to engage with science and consider careers in science, technology, engineering, and mathematics (STEM). Additionally, exposure to Earth and space science research (e.g., via undergraduate research experiences), and subsequent training and mentoring at the PhD and post-PhD levels to develop mission-relevant leadership skills, position individuals on paths to pursue mission leadership should they desire to.

In the course of its work, the committee identified a number of ways in which NASA could make the opportunities to lead missions more accessible, inclusive, and equitable, from making the proposal process more transparent and including DEIA as evaluation criteria, to NASA making consistent and intentional investments in education-career STEM pathways. Moreover, the committee determined that there were valuable lessons to be learned from other federal agencies.

The committee makes 15 recommendations that outline steps or activities that fall within NASA's sphere of influence. The first of which reflects the need for change at the agency/organizational level alongside the specific recommendation actions within NASA SMD to fully realize DEIA in the context of competed missions. The remaining recommendations fall under four general themes: (1) Proposal Process and Review-Enhancing Opportunities, Mitigating Barriers; (2) Data System-Collection, Analysis, and Reporting Data on Proposal Teams, Proposal Pools, and Workforce; (3) Training and Mentoring Potential PIs; and (4) Investment in Career Pathways for Underrepresented Groups. These themes, and selected recommendations that best illustrate each theme, are included below to achieve brevity while also maintaining a comprehensive overview of the report's main messages. Additionally, the recommendations are listed according to the timeframe for their implementation—from near term to long term. Chapter 7 of the report contains the full set of recommendations, and the corresponding recommendation numbers below refer to their order in that chapter. Since competed missions are managed by four divisions of the SMD—Heliophysics, Astrophysics, Planetary Science, and Earth Science—the recommendations are aimed at these divisions. Two other SMD divisions—Biological and Physical Sciences and Joint Agency Satellite Division—do not generate satellite missions. At the same time, most of the recommendations could also apply to all areas of NASA's research, and if the suggested actions were carried out throughout the agency, it would lead to a more diverse workforce across NASA-funded research and training programs.

## OVERSIGHT OF DEIA IMPLEMENTATION AT NASA

There are a number of promising practices used at other federal agencies to address DEIA goals. These include the use of internal and external independent committees charged with promoting and ensuring accountability for the implementation of stated DEIA policies. At the very highest level, the lack of a NASA Advisory Council (NAC) committee focused on DEIA misses a unique opportunity to both help set the tone at the top of the agency as well as ensure ongoing and critical focus on shaping NASA SMD's broadening participation efforts in space missions.

**RECOMMENDATION 1: NASA should empanel an ongoing NASA Advisory Council (NAC) committee specifically focused on diversity, equity, inclusion, and accessibility (DEIA), whose committee chair serves directly on the NAC. This committee should have a broad charter and external world-class membership in this area to directly advise top NASA leadership and ensure an ongoing strong focus on NASA's broadening DEIA efforts.**

## THEME 1: PROPOSAL PROCESS AND REVIEW— ENHANCING OPPORTUNITIES, MITIGATING BARRIERS

NASA's competed space missions are selected from proposals that are submitted in response to a public announcement of opportunity (AO). The processes of mission team formation and concept development, which often takes place one or more years before an AO is made public by NASA, require significant resources from proposers and their institutions. This stage is often referred to as "the competition before the competition" and is idiosyncratic, organic, opaque, and often personality-driven. These informal and organic processes, influenced by "who knows whom" in the community, directly impact the diversity of the PI candidate pool.

For example, many proposals for competed missions are submitted by NASA centers. Thus, NASA is a leading force in the production and determination of the demographic composition of mission PIs. The process of PI selection for NASA center-led competed missions could be much more transparent and committed to supporting traditionally underrepresented PIs. A restructuring of this type would provide a roadmap for other proposing institutions to follow in their own PI selection processes. Moreover, across the range of institutions submitting proposals, the lack of uniformity in non-NASA, internal resources available for preparation and production of mission concepts, full proposals, and site visits may result in a disadvantaged position for some institutions compared to others.

**RECOMMENDATION 2: NASA should work to make the pre-proposal "competition before the competition" process transparent and accessible. Additionally, NASA should use its own resources to expand support of pre-proposal and proposal efforts of diverse, external principal investigators through its field centers and encourage other institutions in the business of supporting and investing in Science Mission Directorate proposals and missions to do the same.**

**RECOMMENDATION 3: NASA should reconsider the requirements for site visits to emphasize the evaluation of technical and programmatic readiness, and eliminate any unnecessary elements. NASA should evaluate the benefit of providing uniform funding to each team that is preparing a site visit, disallowing supplemental funding and other contributions that may result in inequities across teams.**

The steps of the NASA AO process for competed mission proposals are governed by detailed regulations and has become more transparent and uniform across NASA SMD divisions over the past decade. NASA has also taken steps to require and evaluate DEIA plans as part of its AOs for competed missions. However, a comparison with proposal review practices at agencies such as the National Institutes of Health (NIH) and the National Science Foundation (NSF) suggests there is room for scrutiny for sources of bias throughout the NASA AO process. Furthermore, the fact that bias review and DEIA criteria are not currently used in a consistent and systematic fashion results in a limited ability to identify barriers in the mission proposal process, and consequently the specific kinds of interventions that are needed to eliminate them and make progress toward the stated goal of effectively increasing diversity, equity, inclusion and accessibility in the leadership of competed missions. Recommendation 4 outlines a process for tracking, identifying, and reducing bias in the review process for all proposals submitted for research funding.

**RECOMMENDATION 5: In keeping step with its core values of diversity, equity, inclusion, and accessibility (DEIA), NASA Science Mission Directorate should:**

- **Require announcements of opportunity to include a description of how the dimensions of DEIA, including talent development and workforce diversity, are critical to NASA, and require proposals to include a plan for DEIA in the proposed missions describing how the proposed DEIA activities are key for mission success.**
- **Establish a separate, scorable evaluation criterion of the proposed DEIA plans as part of mission proposal review and provide training for reviewers to better equip them to appropriately evaluate proposals concerning DEIA dimensions.**
- **Engage with DEIA experts to implement the new requirements for mission proposals in ways that advance the overarching goal of broadening participation of underrepresented groups in missions.**
- **Establish monitoring and assessment processes to continually measure the impact of the new DEIA plan requirement on progress toward NASA's stated DEIA goals.**

Recommendation 6 provides further ways that DEIA should be enhanced in the form of accountability measures for monitoring and assessing the execution of DEIA plans throughout the mission life cycle.

## THEME 2: DATA SYSTEM—COLLECTION, ANALYSIS, AND REPORTING OF DATA ON PROPOSAL TEAMS, PROPOSER POOLS, AND WORKFORCE

Inadequate data gathering, monitoring, and reporting are key barriers to NASA's understanding of the demographics of its proposer pool and to its ability to measure improvements. Specifically, the inadequate infrastructure for collecting demographic data on teams submitting mission proposals severely limits NASA SMD's ability to accurately determine the relationship between the current mission proposal process and diversity in the pool of proposal leadership teams. For example, further study is needed to evaluate to what extent variations over time and differences between SMD divisions in the proportion of women among mission PI proposers compared to the proportion of women among doctorate recipients in the physical sciences are due to the mismatch between the career-stage distribution of women in these various populations and the current seniority expectation for mission PIs. The current data gathered by NASA on the demographics of proposal teams, for both competed missions and grants, is also inadequate to inform NASA about the diversity in participation of Earth and space scientists from different demographic groups in research and activities funded by the agency. Furthermore, the promise of NASA's existing best practices to address DEIA, both within the agency and PI-led mission opportunities, is limited by the lack of defined metrics to evaluate effectiveness and track progress resulting from these interventions.

**RECOMMENDATION 7: NASA Headquarters (HQ) should develop a systematic and transparent process that employs routine monitoring and tracking of proposal submissions and selections, and submit an annual report of these data to the diversity, equity, inclusion, and accessibility committee of the NASA Advisory Council as well as make the report publicly available. This report should include data on dimensions such as funding rates and diversity in team participation in principal investigator–led missions as well as research and analysis grants; but could also include data on other important dimensions of interest to the agency. NASA HQ should seek professional statistical expertise to set in place the needed infrastructure to support robust data collection, monitoring, and reporting including, but not limited to, adequate staffing, data collection standards of practice, monitoring and analytic systems, annual reporting capability, and external partnerships, to overcome the challenges of tracking participation in NASA's Earth and space science activities.**

Recommendation 8 addresses specific ways NASA should enhance its current data-gathering process, but the issue of the absence of data is not limited to NASA proposals. There is currently no strategy and methodology being employed to effectively gather demographic data on participation in the Earth and space sciences along the whole career pathway—from undergraduate and graduate programs to academic departments and the professional workforce. While NSF and the American Institute of Physics regularly gather information about academic departments (such as numbers and types of bachelor's degrees, doctorate degrees, and faculty in different science fields), the workforce in the Earth and space sciences is usually categorized according to the type of object observed and therefore does not align directly with the disciplinary categories typically represented in such studies. Not only are the SMD divisions interdisciplinary in nature (e.g., Planetary Science comprises of geology, biology, atmospheric science, space physics, etc.), but there is also considerable overlap in the science covered by different divisions. Currently, both the total size of the U.S. post-PhD workforce in each of the disciplines associated with the four divisions of SMD with competed missions and the degree of the overlap between those disciplines are very poorly quantified.

**RECOMMENDATION 9: To regularly assess the state of the profession, NASA Science Mission Directorate should provide funding for professional organizations (e.g., American Institute of Physics, American Astronomical Society, American Geophysical Union, American Physical Society, etc.) to employ the necessary professional expertise in survey methodology and statistical analysis to systematically carry out surveys of the workforce, within and across the four science divisions with competed missions, to inform NASA of the participation of different demographic groups as well as the barriers and opportunities for advancement along entire career pathways in the Earth and space sciences.**

## THEME 3: TRAINING AND MENTORING POTENTIAL PIs

Preparation for competed mission leadership starts early in an Earth and space science career. Training and mentorship opportunities are valuable tools to develop a diverse pool of future mission leaders. Acquiring the necessary information about proposing and navigating the informal pre-proposal process as described above in theme1 presents critical challenges to the aspiring PI, and especially PIs from small and less-resourced institutions. A small number of individuals and institutions are effectively "gate keepers" for selection of mission PIs, and non-PhD granting academic institutions, Historically Black Colleges and Universities (HBCUs) and other

Minority-Serving Institutions (MSIs), where a high concentration of women physics faculty and Black and Latinx physics faculty are employed, experience inequitable access to opportunities that are helpful to scientists who go on to submit competed mission proposals.

**RECOMMENDATION 10: NASA should expand and increase the frequency of training programs that are aimed at encouraging women and historically minoritized communities to become more involved in mission leadership.**

Proposal development, preparation, and submission, including site visits, are resource-intensive processes. Given the concentration of women physics faculty and Black and Latinx physics faculty in less-resourced institutions, such resource-intensive processes usually disadvantage prospective PIs from these underrepresented populations. Diversifying the PI population will require investment in a range of institutions, particularly MSIs. Mentoring and access to collaborative networks that include experienced PIs are formative in preparing those who submit competed mission proposals. However, women and racially minoritized space scientists report less access to mentors and a lower quality of relationship with their doctoral advisors and senior colleagues. Inequitable access to high-quality mentoring relationships is a barrier to increasing diversity in the leadership of competed space missions. Recommendation 11 specifies ways for NASA to provide mission-related research, mentorship, and training opportunities at all stages of professional talent development (i.e., from first-year of undergraduate training through the profession).

**RECOMMENDATION 12: Principal investigator (PI)-led missions present opportunities for aspiring PIs to gain invaluable experiences. NASA should expand resources (e.g., instructional materials, seminars, workshops) for aspiring PIs to gain leadership experience and connect with individuals with mission experience for mentorship opportunities. This may include:**

- **Integrating aspiring PIs as mentees in roles on mission teams, including the higher leadership positions. This could be achieved by including developmental positions in all missions (i.e., competed, non-competed, and instrument investigations), which may require increasing the PI-managed costs.**
- **Encouraging aspiring PIs to pursue entry points to mission leadership, such as proposing to smaller, low-cost mission opportunities (e.g., suborbital, SmallSats and CubeSats, instrument development, and hosted payloads).**
- **Expanding structured networking opportunities at relevant disciplinary conferences such as meet-and-greets where aspiring PIs can connect with collaborators and meet existing PIs, and participate in presentations and question and answer sessions led by NASA personnel about the proposal process.**

Recommendation 13 highlights the need for a broader set of skills and expertise beyond scientific competencies to be considered by NASA in PI training and evaluation.

## THEME 4: INVESTMENT IN CAREER PATHWAYS FOR UNDERREPRESENTED GROUPS

The very low overall retention (~11%) in space science disciplines during undergraduate training, and the accompanying racial/ethnic disparity, is currently a major "pinch point" that restricts the size and diversity of the pool of PhD scientists for future NASA mission leadership. Interventions to improve the diversity and inclusion of STEM students—the ultimate source of proposer pools—are needed at the high school and college levels, where studies show that much of the diversity in relevant fields is lost. NASA needs long-term, sustained investment in effective activities that inspire, educate, train, and mentor, to ensure that the currently small pool of scientists of color have every opportunity to engage in NASA mission-related work and leadership. NASA is uniquely positioned to provide individuals with research experiences that are relevant to future mission roles, and there is arguably nothing more authentic than experiences linked to actual NASA missions.

Informed by the existing knowledge base and recognized exemplar interventions, several guiding principles were identified to inform NASA's investment in these career pathways. The committee found it more important to recommend an intentional strategy through guiding principles to inform NASA's future investments for bolstering STEM pathways rather than prescribing specific interventions. The guiding principles include (1) promoting the development of identities; (2) establishing flexible and relevant STEM education-to-career pathways; (3) intentionally recruiting from historically underrepresented groups; (4) providing access to diverse mentorship; (5) fostering career-life balance; and (6) promoting systemic change around diversity, equity, and inclusion, and are discussed in depth in Chapter 6.

Lastly, the historical underinvestment in the space science research and development infrastructure at HBCUs and other MSIs has in effect limited the capacity at these institutions to compete for a NASA space mission. Moreover, given the racialized patterns of where today's faculty are employed and where degree earners in SMD-related fields are educated, historical underinvestment leads to a disproportionately negative impact on faculty members of color in SMD-related fields and limits access to early-career mission experiences for students of color at MSIs. A previous NASA SMD-funded program, the Minority University and College Education and Research Partnership Initiative, did afford students with training in state-of-the-art research methods and preparation for future NASA leadership, specifically through partnerships with MSIs, but it has been defunded. Currently, the Minority University Research and Education Project (MUREP) and MUREP Institutional Research Opportunity through NASA's Office of STEM Engagement (OSTEM) provide specific engagement to underrepresented and underserved communities in STEM and develop the research capacity and infrastructure of MSIs in areas of strategic importance and value to NASA's mission and national priorities. Further partnership among NASA SMD, OSTEM, and MSIs, leveraging NASA's unique assets, and making NASA-led opportunities more perceptible to individuals and organizations would strengthen participation of underrepresented groups in missions. Likewise, NASA's current investments in STEM education and STEM-related training programs and activities in the United States appear to be concentrated during the K-graduate years of STEM pathways, with very little intentional interventions employed during the post-PhD stages, where attrition continues to be observed. Thus, NASA's influence on the growth and development of a more diverse and inclusive Earth and space sciences workforce is potentially minimized by limited investments along the career path at the post-PhD stage.

**RECOMMENDATION 14: In order to ensure a vibrant, next generation pool of excellent and diverse talent for leadership in competed space missions, NASA Science Mission Directorate, in collaboration with the Office of STEM Engagement, should provide consistent and adequate funding for STEM initiatives that are explicitly centered on diversity, equity, inclusion, and accessibility, address recruitment and retention challenges in the Earth and space sciences, and support and expand opportunities for individuals from underrepresented groups. These investments should reflect a pathways approach spanning the academic and career continuum from post-secondary through post-PhD years in order to establish flexible and robust education-to-career trajectories into the Earth and space sciences workforce, and ultimately into principal investigator–led missions. A systematic process should also be in place to document measurable impacts of these investments.**

**RECOMMENDATION 15: Recognizing the critical role that Historically Black Colleges and Universities (HBCUs), Hispanic Serving Institutions (HSIs), and other Minority-Serving Institutions (MSIs) play in educating and employing women and racially minoritized populations in the Earth and space sciences workforce, NASA leadership, specifically the Science Mission Directorate (SMD) and Office of STEM Engagement Associate Administrators, should charter a joint team to examine and strengthen the historic and current relationship between the two organizations with respect to investments in MSIs. NASA's investments should also redress the historical inequities in NASA supported research and training at these institutions. Specifically, NASA should:**

- **Reinvest in talent development programs in partnership with MSIs specifically related to NASA missions, such as the undergraduate-to-graduate "bridge" type programs previously supported by NASA SMD's Minority University and College Education and Research Partnership Initiative program.**
- **Further leverage NASA programmatic assets such as Minority University Research and Education Programs (MUREP), MUREP Institutional Research Opportunity, and Science Activation to advance broad access to all of its missions, and further enhance early preparation and research engagement of students and early-career researchers, including underrepresented communities.**
- **Provide funding to support mission-related work and activities (e.g., building and designing instruments for spaceflight, hosting science team meetings, etc.) as a means of enhancing research capacity at HBCUs, HSIs, and other MSIs.**

# 1

# Introduction

The National Aeronautics and Space Administration (NASA) has remained committed to ensuring "preeminence of the U.S. in space exploration" as first articulated in the NASA Space Act of 1958.[1] Record numbers of robotic and human spaceflight missions have been launched over the last six decades and the data generated from these endeavors expand knowledge of the Earth, the solar system, and the universe beyond. In today's context of rapidly expanding technology and grand societal and economic challenges, upholding the goal of preeminence is arguably linked to building and maintaining a workforce that is highly skilled, innovative, and productive.

NASA's Science Plan 2020-2024 states,

> As research has shown, diversity is a key driver of innovation and more diverse organizations are more innovative. . . . We will increase support by actively encouraging students and early career researchers. . . . We will also increase partnerships across institutions to provide additional opportunities for engagement and increasing diversity of thought. NASA believes in the importance of diverse and inclusive teams to tackle strategic problems and maximize scientific return. (NASA 2020b)

While NASA is applauded for recognizing the value of diversity and inclusion, and seeking ways to advance these areas in their workforce, even a cursory examination of the teams involved in missions that explore the science of space lays bare the lack of diversity among mission leadership. The realization that mission principal investigators (PIs) and teams are largely white and male provoked this study to identify what NASA could do to increase the diversity of mission teams and their leadership. This section of the report summarizes the nature of the issues with which the committee was tasked and the approach taken to address them. The committee strived to create recommendations of actions for NASA to take immediately and in the long term in order to make significant strides toward its stated goals.

## PI-LED SPACE MISSIONS

The specifics of the early process of selecting missions and the instruments they carried are not very clear in the historical records. It is known that by 1960, the teams of engineers and scientists carrying out missions were selected by NASA. The process for selection seemed to have evolved from first asking specific groups to propose

[1] National Aeronautics and Space Act of 1958, https://history.nasa.gov/spaceact-legishistory.pdf.

instruments on NASA missions (e.g., Pioneer, Venus, Viking), to making a public announcement of opportunity (AO)[2] to propose an instrument to be put on a spacecraft that would be built at a NASA center (e.g., Pioneer, Voyager, Galileo, Cassini, Hubble Space Telescope). In 1988, the first AO (SMEX [Small Explorer]) was released by NASA for a fully open, competed mission led from the outset by a PI, although NASA maintained responsibility for the management and engineering of all segments of the mission except for the instruments and science operations. Additionally, this AO did not include current cost-cap parameters that were implemented beginning in 1994. Since that time, several missions although not fully open, competed missions, were catalysts for the PI-led mission process as we know it today, such as the Near Earth Asteroid Rendezvous (NEAR), the Far Ultraviolet Spectroscopic Explorer (FUSE), and New Horizons that were subsequently "grandfathered" into the Discovery, MIDEX (Medium-Class Explorer), and New Frontiers programs, respectively. Dates of some major milestones in the evolution of NASA's PI-led space missions are listed in Box 1.1.

A PI-led mission is a complete cost-capped, focused, space or Earth science mission in which the concept, including design, development, launch, operations, and scientific analysis, is proposed by a single PI (NRC 2006). Over the years, enthusiasm for PI-led, competed instruments and missions has waxed and waned at NASA Science Mission Directorate (SMD). But there is general recognition that open competition leads to innovation and ingenuity, while having a cost-capped mission limits the growth in the net mission price.[3] At the same time, there remain larger, scientific, strategic purposes for large "flagship" missions led by a NASA center, including but not limited to operating an evolving science program that supports iteration as the mission proceeds, as opposed to a more fixed science program, and a need to accomplish a broad suite of objectives rather than a single or small number of tightly relevant objectives (NASEM 2017a).

NASA's competed missions may be broadly grouped by size, typically within three cost ranges: small size, low-cost missions (<$250 million fiscal year [FY] 2019); medium-size, medium-cost missions ($250 million to $650 million FY2019); and large-size, high-cost missions (>$650 million FY2019). There are also what are called missions of opportunity (MoOs), which are investigations characterized by being part of a non-NASA space mission of any size (either in the United States or an international agency) and having a total NASA cost of under $55 million (FY2019). Competed instruments are a third type of mission opportunity that involve state-of-the-art scientific instruments to be flown as part of a larger mission for the purpose of acquiring information (e.g., images, spectra, in situ particles, and fields, etc.), and the cost of these opportunities span the low-cost to medium-cost mission ranges. Between 1996 and 2019, the total number of competed missions that were proposed is 701, of which 143 (20%) were selected for flight. Over the same time period, the net number of competed instruments that were proposed is 232, of which 46 (20%) were selected for flight. In this report, the review of competed space missions is limited to the last decade (2010-present), in order to focus on factors and issues most relevant to the current era.

With respect to the demographic makeup of the pool of PIs of competed mission proposals, a total of 198 proposals were submitted to NASA SMD between 2010-2019, of which 16% had women PIs. In this same time period, 46 proposals were selected (23%), of which 28% had a woman PI. It is important to note that these data reflect (inferred) gender of PIs, which characterizes the nature of the data currently collected by NASA. Additionally, data on the racial/ethnic background of mission proposal PIs was not available for the committee to consider in this study (see Chapter 3 for a more in-depth discussion of the demographics of competed space mission PIs and teams).

The role of a PI on a competed mission is substantial. The job requires considerable experience in the scientific research field, knowledge of mission design, building of hardware and concepts of mission operation, plus ability to lead and manage a large team of scientists, engineers, and administrators. Currently, PIs tend to be at senior levels of their career (e.g., >20 years since acquiring a PhD), and thus, the demographics of mission PIs most likely reflect the population at such senior levels of career. Consequently, one needs to consider the full pathway in the Earth and space sciences, starting with inspiring young people to engage with science and consider science, technology, engineering, and mathematics (STEM) careers, and following with exposure to space science research (e.g., via undergraduate research experiences), and subsequent training and mentoring at PhD and post-PhD levels, encouraging potential PIs to develop the skills needed to lead a competed mission. While steps are certainly needed

[2] See Appendix A for a list of acronyms.

**BOX 1.1**
**Some Milestones of Principal Investigator–Led Space Missions**

| | |
|---|---|
| 1965 | First African American instrument principal investigator (PI) on sounding rockets and satellites X-ray solar telescopes, PI Art Walker |
| 1972 | First African American instrument PI on the MoonvEUV telescope, PI George Curruthers |
| 1970[a] | Competed instruments selected for Pioneer 10 and Pioneer 11 |
| 1972 | Competed instruments selected for Voyager 1 and Voyager 2 |
| 1976 | First Planetary Science competed instrument with woman PI Galileo Magnetometer, PI Margaret Kivelson |
| 1988[b] | First competed Small Explorer (SMEX) mission announcement of opportunity (AO)—small, multidisciplinary<br>SAMPEX, PI Glenn Mason—Heliophysics<br>FAST, PI Charles Carlson—Heliophysics<br>SWAS, PI Gary Melnick—Astrophysics |
| 1994 | First fully competed mission announcement of opportunity (AO)—Discovery, Planetary Science<br>Lunar Prospector, PI Alan Binder<br>Stardust, PI Robert Brownlee |
| 1995 | First competed Astrophysics instrument with woman PI—launched PCA on Rossi XTE, PI Jean Swank |
| 1997 | First openly competed MIDEX mission AO<br>IMAGE, PI James Burch—Heliophysics<br>WMAP, PI Charles Bennett—Astrophysics |
| 1997 | First competed Earth System Science Pathfinder (ESSP) launched GRACE, PI Byron Tapley |
| 1999 | Competed instruments on Earth Observing System launched |
| 2002 | Competed instrument on James Webb Space Telescope with woman PI NIRCam, PI Marcia Rieke |
| 2002 | First openly competed New Frontiers mission selected Juno, PI Scott Bolton |
| 2005 | First Astrophysics competed SMEX missions with a woman PI selected NuStar, PI Fiona Harrison<br>GEMS, PI Jean Swank |
| 2005 | First mission of opportunity (MoO) instrument with a woman PI selected Moon Mineralogy Mapper, PI Carle Pieters |
| 2007 | First Planetary Science competed Discovery mission with a woman PI selected GRAIL, PI Maria Zuber—Planetary Science |
| 2017 | First New Frontiers mission with a woman PI selected—Large, Planetary Science Dragonfly, PI Elizabeth Turtle |
| 2021 | First Earth Science (EVM) competed mission with a woman PI selected INCUS, PI Susan van den Heever |

[a] Degree of openness of competition not clear.

[b] NASA responsible for management and engineering of all segments of mission except instruments and science operations.

in the short term to bolster diversity, equity, inclusion, and accessibility (DEIA) in the leadership of competed space missions, fully realizing them in the mission PI landscape is a multi-generational challenge.

## CHARGE TO THE COMMITTEE

In order to address its stated goal of enhancing DEIA in the leadership of space missions, NASA requested that the Space Studies Board (SSB) in collaboration with the Board on Science Education at the National Academies of Sciences, Engineering, and Medicine establish the Committee on Increasing Diversity and Inclusion in the Leadership of Competed Space Missions to conduct a new study of PI-led missions in the space sciences, with respect to diversity within the pool of proposal leaders. National Academies staff recognized that to address the task, the expertise of both space scientists with mission experience as well as social scientists would be needed. To this end, the 14-member expert committee reflects a wide range of disciplinary fields with broad experiences in academia, government agencies, and other institutions involved in space sciences (Appendix D includes brief biographies of the committee members and staff).

Whereas previous SSB committees considered aspects of PI-led missions such as issues affecting cost and schedule (NRC 2006), this committee was specifically tasked with recommending actions to increase DEIA in the leadership of space mission proposals submitted to NASA SMD's competed space mission programs (see the committee's charge in Box 1.2). In this report, the committee articulates the end-to-end proposal process for competed space missions and offers recommendations for enhancing DEIA within proposed leadership teams.

## STUDY APPROACH

Over the course of this study, the committee held eight fact-finding meetings, heard presentations by past and present representatives of NASA, research institutions, universities, professional societies, as well as social science scholars and representatives from other federal agencies. The topics covered at the meetings included the space sciences workforce demographics; mission proposal development, review and selection; NASA-collected demographic data on proposers; institutional and organizational processes for forming mission teams; policies, procedures, and programming across federal agencies for advancing DEIA in science and engineering (S&E) fields; and social factors impacting DEIA of mission teams, such as intersectionality and interdisciplinarity (see the Preface for the list of all presenters). The committee and study were set up during the COVID-19 pandemic, so the study was an entirely virtual process. The first virtual meeting was held on January 20, 2021, followed by a total of 11 meetings, with the last meeting held November 30 and December 1, 2021.

Unlike traditional consensus studies at the National Academies, the work of this committee primarily revolved around assessing NASA SMD's mission proposal process by applying committee members' collective expertise to analyze different facets of the current process in theory and in practice. In this report, some of the evidence considered by the committee is not published in peer-reviewed journals, but rather is the physical and oral documentation of aspects of the process and demographics on proposers as provided to the committee by NASA and representatives of organizations involved in space missions at NASA centers, universities, industry, and research institutes. The committee also leveraged the extensive body of research in the social sciences on equity, inclusion and bias in scientific organizations, and on pathways in STEM education and careers.

Addressing the study charge required that the committee identify and understand barriers to diversity that may exist within the mission proposal process itself. The committee relied in part on public testimony and on the personal experiences of committee members. The committee also drew on qualitative data collected via structured interviews with competed mission PIs in consultation with NORC at the University of Chicago and the findings (see Appendix C) informed the drafting of the report and the committee's conclusions and recommendations.

**BOX 1.2**
**Charge to the Committee**

The National Academies of Sciences, Engineering, and Medicine will convene an ad hoc committee to address the topic "Increasing Diversity and Inclusion in the Leadership of Competed Space Missions." This committee will recommend actions to increase diversity, inclusion, equity and accessibility in the leadership of space mission proposals submitted to the NASA Science Mission Directorate (SMD) competed space mission programs. To accomplish this goal the committee will:

1. Map the current space mission proposal system. A system mapping will lay out requisite tasks, organizational processes, and human networks necessary to navigate the competitive process. The system mapping will also provide the SMD-supported science community with a useful reference for developing and submitting a space mission proposal.
2. Identify elements of the system that may present barriers or bottlenecks to potential proposers, yielding a less diverse pool of proposal leadership teams. As part of this step, the committee will identify points in the system at which NASA actions may have large, positive effects as well as acknowledging those points in the system at which the agency can do little to affect the diversity and inclusiveness of proposal leadership teams.
3. In the step above, identify humanistic elements of the system that may present impediments to applicants, limiting the diversity of the competitive pool. Such elements include, but are not limited to work/life balance considerations, professional advancement considerations (e.g., requirements for tenure), institutional support, the impacts of explicit and implicit biases and the impacts of discrimination, harassment, lack of inclusion and lack of accommodation for disabilities and any differences between the communities associated with different SMD divisions. In performing this comparison, the committee may engage with survey and demography experts.
4. Use existing NASA proposer data and other sources to analyze differences between the pool of current and past SMD proposers and the community of researchers in NASA-relevant fields, the larger U.S. science community and the national population. In performing these comparisons, the committee may engage with appropriate, relevant experts and may conduct interviews with relevant organizations and federal agencies if needed. Where significant, identify differences from the overall pool that are specific to the communities of SMD's four science divisions.
5. In developing recommendations, consider best practices from other agencies funding large, PI-led, multi-institutional projects to find common bottlenecks and to identify innovative methods for overcoming such barriers. Example programs that may be examined include NSF's Engineering Research Centers and Material Science Research Centers and NIH's "program project" grants.
6. Recommend practical and effective actions that NASA SMD and its partners can execute to help proposers overcome barriers to diversity and inclusion to ensure a diverse pool of proposal leadership teams.
7. Recommend ranges of activities and choices that individuals may make during their career to enhance the likelihood of individual success as a future leader of a competed space mission. These recommendations should balance activities against bias and barriers.

As part of its work, the committee was also asked to review and consider findings of numerous past social science studies of participation barriers for women and minorities in professional endeavors such as science and engineering.

Qualitative research involves the systematic examination of social phenomena from the perspective of those individuals who experience them (Guba and Lincoln 1994). Through rigorous data collection and analytical techniques, qualitative research aims to understand phenomena "in terms of the meanings people bring to them" (Denizen and Lincoln 2005). In the context of this study, the committee was interested in understanding how individuals experienced the mission proposal development, preparation, and submission process, including any barriers that they may have encountered before, during, and after the process. Given that the proposal process is a social process and not just a technical one, qualitative methods provided the means to gain an in-depth understanding of how proposers made meaning of these barriers in the context of their identity and previous educational and professional experiences, as well as how they navigated any such barriers. The committee identified qualitative interviewing as the most appropriate method given the research questions that emerged from the study charge and the committee's review of the relevant literature. Interviewing allowed for comparison across subjects and elucidated the internal motivations and feelings that may drive or result from observable behaviors (Lamont and Swidler 2014). Though not without limitations, the structured interviews allowed the committee to enrich its understanding of how the mission proposal process proceeds in practice. The committee's decision to only focus on PIs within the context of this qualitative study was motivated by (1) the need to maximize the probability of acquiring a clear, defensible sampling in a very small, pilot study (n <40) and the role of PI is a part of every mission leadership team and (2) the need to identify themes that with reasonable confidence could be attributed to proposal development and submission. This qualitative study represents an important first step, but the committee acknowledges the need for a much larger study in the future that can accommodate the variety of roles and perspectives on mission teams and of aspiring PIs, as well as accommodate sufficient participation to achieve statistical significance. Although the qualitative study population includes those who were ultimately successful at submitting at least one proposal as PI, many also offered their reflections on past unsuccessful experiences.

When reaching conclusions and developing recommendations, the committee drew on multiple streams of evidence, and in particular complemented oral testimonies with evidence from relevant research literatures. When oral testimony was the sole basis of the finding, the committee took care not to present the finding as supported with adequate evidence, but instead framed them as potential issues for NASA SMD to consider and investigate further, without making recommendations for a specific course of action. As a result of these deliberative processes, all conclusions and recommendations outlined in this report, reflect the full consensus of the committee.

Given the committee is tasked to recommend ways to increase diversity and inclusion in the leadership of competed space missions, the committee also benefited from discussions on the various ways that the terms "diversity" and "inclusion," as well as other terms like "equity," "accessibility," and "intersectionality" have been used in the research literature to determine how they would be applied in this report. The committee recognizes that there is not consensus in the field around these terms, and so Box 1.3 reflects this committee's adoption of definitions of particular terms that guided how these aspects are discussed in this report.

An important factor that the committee gave consideration to in this report is the makeup of the U.S. space science workforce (i.e., astrophysics, Earth sciences, heliophysics, and planetary sciences) and, in particular, the fraction of the workforce comprising foreign-born individuals. For example, data from the American Institute of Physics show that for the past ~30 years, about half of the PhDs in physics were awarded by U.S. universities to foreign-born scientists. Many such students stay in the United States pursuing research careers. Additionally, non-U.S. scientists immigrate as post-doctoral researchers. Thus, a significant fraction of the space science workforce are foreign-born scientists and engineers who make substantial contributions to NASA's space missions. They also contribute to the diversity of the space sciences workforce, with respect to gender and race/ethnicity.

In this report, however, the focus is placed on the overall domestic S&E workforce for two reasons:

1. It is important to provide opportunities for every U.S.-born person to pursue a career in the space sciences should they want to, particularly increasing the opportunities for individuals from underrepresented communities (NASEM 2019a; Ochoa and McCrary 2021).
2. It is not clear how the supply of foreign talent will change in the near future due to the ebbs and flows with national policies and with demographic and economic realities of source countries (AAAS 2021; Khanna 2021; NSB 2022).

Moreover, the changing demographics in the United States suggests that within 40 years, no single racial or ethnic group will comprise the "majority" population group in the United States, demonstrating potential opportunities to diversify the future space science workforce and to cultivate the next generation of U.S.-born mission leaders. The committee interpreted the charge to "identify humanistic elements of the system that may present impediments to applicants" as requiring an assessment of the full scope of career pathways to PI leadership so that actionable recommendations to improve diversity would be developed.

Lastly, the committee notes that while this study was being carried out, NASA initiated four relevant activities:

1. In early 2021, NASA requested an additional National Academies consensus study, "Foundation for Assessing the Health and Vitality of the NASA Science Mission Directorate's Research Communities," to provide guidance on the data that NASA needs to collect to appropriately analyze its research community's health and vitality and recommend actions to improve the health and vitality of the research community toward realizing NASA research objectives.[3]
2. In June 2021, NASA released a "Request for Information on Advancing Racial Equity and Support for Underserved Communities in NASA Programs, Contracts and Grants Process" to receive public input on NASA's mission directorates' programs, procurements, grants, regulations and policies, with the purpose of improving these activities such that systemic inequitable barriers and challenges facing underserved communities are removed (NASA 2021f).
3. In September 2021, NASA SMD released a "Request for Information on Adding Inclusion, Diversity, Equity, and Accessibility Requirements to NASA's Announcements of Opportunity," announcing the potential inclusion of explicit DEIA evaluation criteria within the mission proposal process, and requesting input from the broader community on the reasonableness and feasibility of this proposed new requirement (NASA 2021g).
4. In September 2021, NASA also released a DEIA policy statement expressing the agency's commitment to being a leader in advancing just, equitable, diverse, accessible, and inclusive NASA-funded workforces and NASA's internal workplaces (NASA 2021e).

The committee commends NASA for enacting a number of efforts that are intentionally focused on DEIA, and it remains to be seen what impact such activities will have on the diversity of the space science workforce and of competed mission leadership.

## ORGANIZATION OF THE REPORT

This report describes the history and current status of the diversity of PI-led mission teams, the ways in which certain practices have affected the diversity of PI-led missions, and the steps that can be taken to help ensure DEIA within future PI-led missions.

In Chapter 2, the committee maps out the steps typically taken to develop and submit a complete mission proposal. The initial process of team formation and concept development starts months to years before NASA announces an opportunity to propose a relevant type of mission. The subsequent process of constructing the proposal depends on substantial institutional support. Along the way there are both opportunities and barriers to success, including the proposal review process where ways of removing any possible sources of bias are explored.

Chapter 3 presents data on the composition of competed mission teams for four science divisions of NASA SMD: Astrophysics, Heliophysics, Planetary Science, and Earth Science. Before considering the means of increasing diversity and inclusion in the leadership of competed space missions, one needs to consider the demographics of the relevant workforce. The demographics gathered by NASA's Solicitation and Proposal Integrated Review and Evaluation System (NSPIRES), the primary mechanism for all proposal submissions, are compared with

[3] National Academies of Sciences, Engineering, and Medicine, "Foundation for Assessing the Health and Vitality of the NASA Science Mission Directorate," https://www.nationalacademies.org/our-work/foundation-for-assessing-the-health-and-vitality-of-the-nasa-science-mission-directorates-research-communities.

**BOX 1.3**
**Glossary**

**Accessibility:** The design of products, devices, services, or environments so that people with disabilities can fully participate, engage, and benefit.

**Diversity:** The practice or state of involvement of persons across a variety of social and demographic characteristics within organizational, institutional, and interactional settings.

**Equity:** The absence of barriers, biases, and obstacles that impede access, fair treatment, and opportunity for contribution by all members of a community, recognizing that different resources or approaches are needed to remedy the uneven playing field that exists across different groups.

**Ethnicity:** Socially constructed and historically contingent categorizations of humans based on perceptions of shared culture, such as language, ancestry, practices, and beliefs.

**Gender identity:** The way one identifies oneself within or outside of traditional, socially constructed categories of women and men. Gender identity is separate from biological sex (which is a socially constructed dichotomy of bodies based on perceptions of shared physical and/or chromosomal traits), and may include gender non-binary and gender queer identities that lie outside of the traditional gender binary.

**Historically minoritized communities in STEM:** Socio-demographic groups (e.g., certain genders, racial/ethnic groups) that, as a result of historical and contemporary processes of oppression and bias, have been excluded from full and representative participation in STEM.

**Inclusion:** The practice of facilitating the equal distribution of opportunities, resources, and recognition to persons regardless of social and demographic characteristics within organizational, institutional, and/or interactional environments.

**Intersectionality:** A critical, theoretical, and analytical framework that highlights how multiple social identities such as race, gender, sexual orientation, socioeconomic status (SES), and disability intersect at the micro level of individual experience to reveal interlocking systems of privilege and oppression (i.e., racism, sexism, heterosexism, classism) at the macro socio-structural level.

**Race:** Socially constructed and historically contingent categorizations of humans based on perceptions of shared physical and/or social traits.

**Sexual identity:** The way one identifies oneself in terms of to whom one is romantically or sexually attracted. May be within or outside of traditional, socially constructed categories of heterosexuality and non-heterosexuality, and may include bisexual, queer, fluid, and asexual identities.

**Underrepresented communities:** Socio-demographic groups (e.g., by gender, race/ethnicity, class status) whose proportional representation within a specific organization, institution, or profession is lower than their representation in the national population overall.

workforce data gleaned from published surveys of the space sciences and from reports on the demographics in S&E fields at the national level.

In examining the lack of diversity in the space sciences workforce, understanding typical career tracks is key. Chapter 4 discusses the state of education-career pathways into the space sciences from the undergraduate level into professional stages and identifies potential opportunities for bolstering the participation of underrepresented communities in competed space missions.

Along education-career pathways into the space sciences, and ultimately into mission PI-ship, both facilitators and barriers of success exist. Chapter 5 describes some of the factors that may hinder the development, preparation, and submission of successful competed mission proposals, as well as diversity in the space sciences workforce more broadly. This chapter also describes the disproportionate impact of various factors on communities that remain underrepresented in the space science workforce and draws on the commissioned qualitative research study of the experiences of a sample of PIs in putting together a competed space mission proposal.

Other science agencies manage large, PI-led multi-institutional projects—the National Science Foundation, the National Institutes of Health, the Department of Energy, the Department of Defense, etc.—and also seek to promote DEIA in these projects and within STEM fields more broadly. Chapter 6 describes some of the evidence-based, promising, and emerging practices that have been employed at the federal level for fostering DEIA that would be useful for NASA to consider toward realizing its stated goal of ensuring a diverse pool of proposal leadership teams.

Finally, Chapter 7 summarizes the core findings and recommendations to NASA, to institutional partners, and to future proposers. The recommendations presented in Chapter 7 directed to NASA are mainly aimed at the four science divisions of SMD—specifically Heliophysics, Astrophysics, Planetary Science, and Earth Science—because the competed missions are managed by these divisions. At the same time, most of the recommendations could also apply to all areas of NASA, and if the suggested actions were carried out throughout the agency, it would lead to a more diverse workforce across NASA.

# 2

# Understanding the Mission Formulation and Proposal Process

In its 2020 Science Strategy, the National Aeronautics and Space Administration (NASA) Science Mission Directorate (SMD) states: "SMD seeks to discover the secrets of the universe, to search for life, and to protect and improve life on Earth" (NASA 2020b). NASA SMD's flight missions are essential to addressing critical scientific needs stated in the NASA's strategic plans, and exist as competed principal investigator (PI)-led missions and very large, "flagship missions" that are led through NASA centers. Specifically, PI-led missions exist in a larger context of a "balanced portfolio" that encompasses a broad range of scientific investigations that includes flight missions, research and analysis, technology development, and applied/fundamental science. These projects together support NASA's strategic plans, and the relative balance across these efforts is informed by the National Academies of Sciences, Engineering, and Medicine through its decadal surveys and is responsive to Administration priorities and direction from Congress.

PI-led missions have been an increasingly important element of the balanced portfolio over the past several decades in the four science divisions in NASA SMD that conduct spaceflight missions: Astrophysics, Earth Science, Heliophysics, and Planetary Science. The fifth division, Biological and Physical Sciences, does not generate satellite missions. Competed opportunities for NASA-funded investigations vary in scope, complexity, and cost from ~tens of millions of dollars for instruments or Small Innovative Missions for Planetary Exploration (SIMPLEx) investigations to over $1 billion for New Frontiers or Astrophysics Probes investigations (see Table 2.1 in the next section). The processes whereby NASA advertises competed opportunities—announcements of opportunity (AOs)—and makes selections based on proposals developed by aspiring PIs have evolved significantly and quite effectively since the 1970s. In its 2006 report *Principal-Investigator-Led Missions in Space Science*, the National Research Council reported that the space science community sees a lot of scientific value in PI-led missions as a complement to NASA flagship missions (NRC 2006).

Investments in a flight investigation are large, requiring contractual vehicles[1] (or other appropriate vehicles for NASA centers or other government entities) that provide enforcement and oversight mechanisms to ensure that the investment is well managed and can deliver on its promise. The PI on each selected proposal is assigned responsibility and authority for the success of the investigation, through a contract between NASA and the PI's institution. The requirements are defined in detail in a NASA Supplement to the Federal Acquisition Regulations (FAR) and are presented in summary form in this report. The 2006 report on PI-led missions broadly examined

[1] A contractual vehicle is the method the government uses to buy goods and services.

PI-led missions, from the process of proposals to selection to the execution of missions (NRC 2006). The recommendations in that report addressed the factors that contributed to the success and challenges of PI-led missions and provided some recommendations related to the selection process, concept studies, selection criteria, team development, funding profiles, International Traffic in Arms Regulations (ITAR) compliance, and termination reviews. The scope of this report is focused on diversity and inclusion in the leadership of competed space missions and will not discuss the execution of missions.

The mission proposal process leading up to selection can be divided into three stages: (1) the development of an idea into a mission concept (concept development); (2) proposal development and submission; and (3) in the case of two-step mission proposals, preparation of a concept study report (CSR) for concepts that are selected through the Step 1 process. It is important to note that proposal preparation for competed missions begins far in advance of an AO released by NASA, typically 1 or more years before the release. Once a proposal has been developed and submitted, proposal evaluation and review of CSRs are almost entirely under NASA's direct control. The overall process, at a high level, is captured in Figure 2.1. NASA follows strict procedures specified in federal regulations, time tables, and rules that are generally transparent and well documented. In contrast, concept development, which has been referred to as "the competition before the competition," is idiosyncratic, organic, opaque, and often personality-driven.

This chapter discusses the current proposal process for competed space missions as understood by this committee through its examination of disparate sources of evidence and testimony from members of the space science community. In the sections to follow, the committee begins with a look at concept development, team formation, and proposal development, and then the NASA proposal evaluation process in more detail.

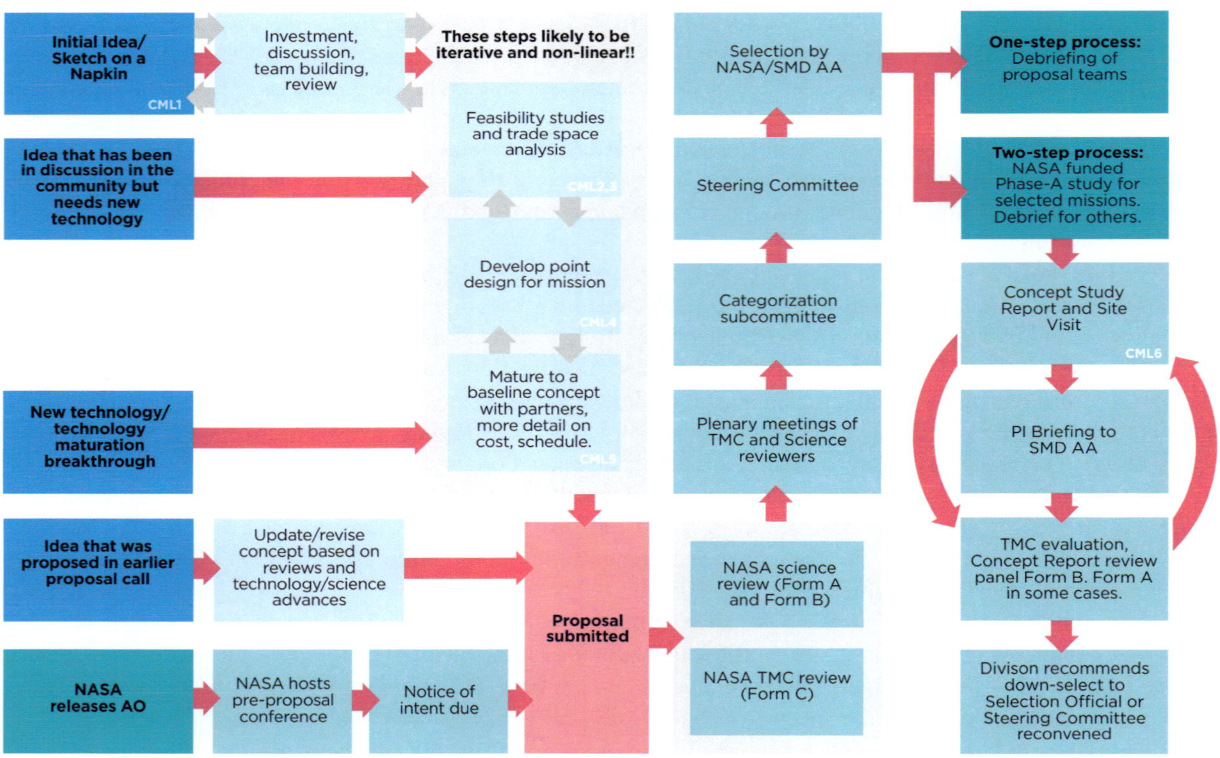


**FIGURE 2.1** Overall flow diagram of the process of developing mission concepts and the NASA proposal evaluation and selection process. NOTE: AA, Associate Administrator; AO, announcement of opportunity; SMD, Science Mission Directorate; TMC, technical, management, and cost.

## FORMING A TEAM AND DEVELOPING A PROPOSAL

### Concept and Technology Development

Concept development, which begins years before a draft AO or other solicitation is released by NASA, is effectively a competition before the competition that starts when a proposal is submitted. In the concept development phase, an aspiring PI formulates and matures a concept for a mission and assembles a team of key co-investigators (Co-Is)—critical scientific partners at other institutions—and supporting institutions that bring a range of expertise to the development and execution of the mission. Teams are formed to develop early-stage ideas into more mature concepts, as well as to build mission designs and prepare complete proposals from mission concept studies. The team formation is intertwined with the concept development, where often a scientist or engineer, or a small group of scientists or engineers, conceive of a new measurement or mission. As the concept becomes more mature, people with needed skills or expertise are brought onto the team. The process of going from a new concept to a mission design often takes many months or years. Many ideas are proposed numerous times before they are selected, so the team may evolve and develop during the iterations. A typical team composition during the early phases of concept development may include science experts, mission designer, project systems engineer (PSE), project manager (PM), instrument lead, and ground data system engineer. During concept development, a team may engage in technology development, and this may be supported by NASA instrument or concept development grants, or it may be supported by institutional investments. In this stage, the team also begins generating proposal material in anticipation of a future AO, and the PI and others may publicize the mission concept at conferences and in published papers and/or make presentations to appropriate NASA and other committees. After the release of an AO, the teams refine and submit the proposal, but the concept development stage can easily stretch over 1 year and often extends over many years and across multiple AOs.

Since the conception of a mission idea, formation of a team, and maturation of a concept to be proposed precede the NASA release of an AO, several frameworks have been conceptualized to map the proposal process from idea to fruition. One example is the concept maturity level (CML) framework (Wessen et al. 2013). The CMLs span from level 1, a new idea sketched on a napkin, to level 8, an actual project (see Figure 2.2), and brief descriptions of CMLs 1 through 6 are as follows:

- *CML 1 Sketch on a Napkin*—The science questions have been well articulated, needed science observations identified, rough sketch of mission concept of operations and objectives, and an assessment of technology readiness level.
- *CML 2 Initial Feasibility*—The idea is expanded and questioned on the basis of feasibility, from a science, technical, technology readiness, and programmatic viewpoint.
- *CML 3 Trade Space*—Exploration done of the science objectives and architectural trades to explore the relationship between science return, cost, and risk.
- *CML 4 Point Design*—A specific design and cost that returns the desired science has been selected within the trade space. Subsystems trades have been performed.
- *CML 5 Baseline Concept*—Implementation approach has been defined including partners, contracting mode, technology maturation, integration and test approach, cost and schedule. Appropriate for Step 1 proposals.
- *CML 6 Integrated Concept*—Expanded details on the technical, management, cost and other elements of the mission concept have been defined and documented. A NASA Step 2 CSR is at this level of maturity.
- *CML 7 and 8*—Correspond to selected or assigned missions in early implementation, preparing for design reviews.

The effort, time, and financial resources to take an idea from CML 1 to CML 6 are significant (see Annex 2.A for more complete details and typical resource requirements to achieve each level). Concepts typically need to mature to at least CML 5, "Baseline Concept," to be further developed into a mission proposal. The concept maturation process can be non-linear, where trade spaces are revisited after a point design is explored, or technology

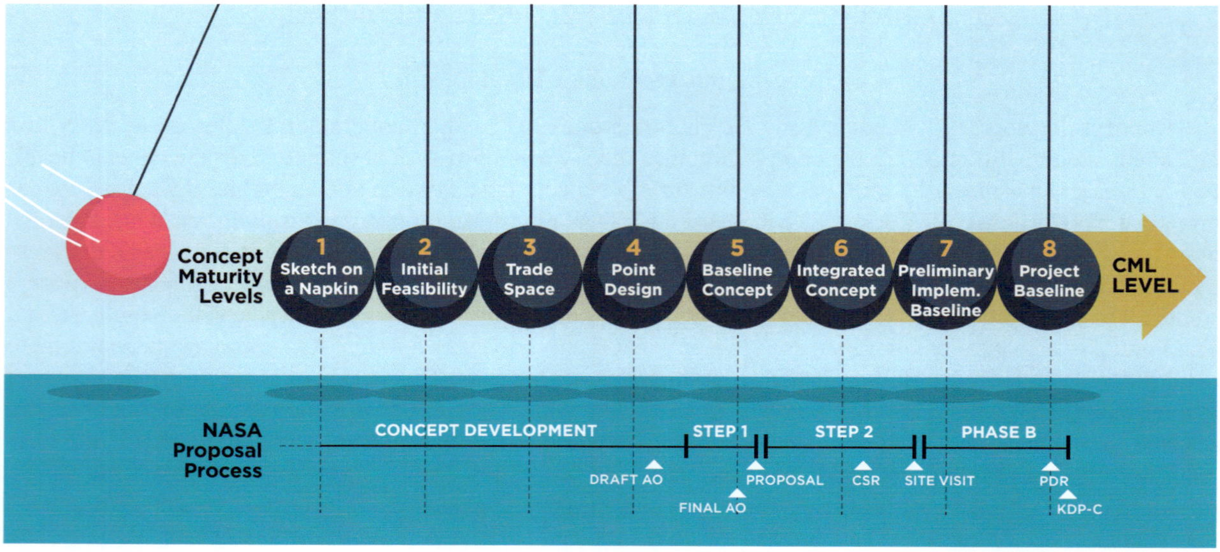


**FIGURE 2.2** The parallels between CMLs and the NASA proposal and early implementation process. NOTE: AO, announcement of opportunity; CML, concept maturity level; CSR, concept study report; KDP, key decision point; PDR, preliminary design review.
SOURCE: Based on concepts from Wessen et al. (2013).

## BOX 2.1
## Summary of NASA Project Phases

**Program Pre-Formulation**

- Pre-Phase A: Concept Studies

**Program Formulation**

- Phase A: Concept and Technology Development
- Phase B: Preliminary Design and Technology Completion

**Program Implementation**

- Phase C: Final Design and Fabrication
- Phase D: System Assembly, Integration and Test, Launch
- Phase E: Operations and Sustainment
- Phase F: Closeout

More on NASA mission phases can be found at https://www.nasa.gov/seh/3-project-life-cycle.

challenges require revisiting the trade space. A great degree of overlap exists between the CMLs and the NASA mission proposal process (see Figure 2.2), the latter beginning after the AO is released and continuing beyond proposal submission. Figure 2.2 also shows the relationship of CMLs to the NASA project phases that encompass the lifetime of a mission. The standard NASA project phases are listed in Box 2.1, although the scope of this report is primarily related the Pre-Phase A through Phase B of the NASA framework.[2]

The steps preceding proposal development include feasibility studies, trade space studies, and full conceptual design (or point design). These steps culminate in several specific mission concept items such as a notional concept of operations plan, an instrument technology readiness level (TRL) discussion, and science traceability including high-level science goals traced down to instrument requirements and then capability. These steps may also include other items like securing a spacecraft provider; plans for subsystem testing, budget and scheduling; and establishing a management structure. Taken together, the progress from CML 1 to CML 5 requires extensive resources—team formation, funding for feasibility and trade space studies, engineering and scientific analysis, and often team reviews as a baseline concept is refined. Thus, concept development alone requires an institutional investment and significant effort by a science team, a core group of engineers, a PSE and PM, and financial and budget analysts. It is clear that a significant amount of work is required prior to the release of an AO by NASA, and this can be thought of as another form of a competition. It is also important to note that while the CMLs depicts a real-world process of how concepts become missions, they are not meant to be an exact description of the steps required to be taken. They are also not a formal NASA scheme like project phases and TRLs, which are discussed later.

Another aspect of development that occurs well before proposals are submitted is technology development. Often new measurement capabilities are enabled by new and advanced technologies in instrumentation. NASA expects the technology included in mission proposals to be sufficiently mature, or at high TRL. NASA's defined TRLs range from 1 to 9, where TRL 1 is technology in an early development stage, and TRL 9 is flight proven through successful mission operations (NASA 2012). In advance of including new technology in proposals, teams may be investing and developing the new technology to move them to the appropriate maturity levels. This is an area where experience and familiarity with the technology development, resources, and pathways may be a challenge to new PIs. There are many NASA proposal opportunities specifically focused on technology development, some with specific TRL development objectives. Funding levels range from $200,000 to $300,000 per year to >$1 million per year.

## Becoming a PI

Becoming a PI or Co-I on a winning proposal is typically a long training and development process for the investigators as well as the proposing team as a whole. The PI's responsibilities range from concept formation and maturation, leadership during proposal development, and if selected, responsibility for the overall execution of the mission. The resources required for this pursuit in terms of dollars and time are substantial, therefore creating a challenging environment for investigators, particularly new ones with many expectations as listed below:

1. Defining, building, and leading the team well prior to any AO;
2. Being aware of AOs, their release and cadence, and recognition that the process starts well in advance of the release of the AO;
3. Developing a compelling scientific investigation concept that is responsive to priority questions in National Academies decadal surveys and other NASA strategic needs;
4. Securing committed collaborations among investigators to refine and pursue the concept;
5. Defining an instrument and/or suite optimized for the pursuit of the investigations;
6. Identifying and securing a partner institution experienced in putting together winning proposals, knowledgeable of policies and requirements, and that is willing and able to invest institutional bid and proposal (B&P) resources in the proposal;

[2] See "NASA Program/Project Life Cycle," Section 3.0 in NASA (2019).

7. Dedicating the time and expertise required to develop, review, and refine a proposal that is compliant, compelling, and timely;
8. Taking ownership and accountability for the investment in the proposal and for the technical and programmatic success of the investigation if selected; and
9. Demonstrating strong perseverance and a commitment to winning.

Mission teams are formed to provide the scientific and technical skills and knowledge to write a winning proposal and implement a successful mission. Individuals in these roles are sourced from NASA centers, industry, nonprofits, universities, and university-affiliated research centers—often a web of partnerships is required to fill all roles effectively. Many PIs or teams submit a number of proposals before winning (Zurbuchen 2019). As teams can form around the socialization of ideas often years before an AO, this could translate to potentially more than a decade spent championing an idea, proposing and re-proposing, before a mission proposal is awarded. This lengthy process limits participation of earlier career scientists and contributes to shaping the pool of potential PIs.

**Finding:** The expectations of the principal investigator of a NASA competed mission are significant and wide ranging, from science expertise to assembling a team, shepherding proposal development, and accepting ownership and accountability for the technical and programmatic success of not just the proposal, but the implementation of the mission.

### Team Formation and Proposal Development

Team formation as part of concept and proposal development involves individuals (PI and Co-Is) and institutions, typically the PI's home institution as host, a managing institution (often but not always a NASA center), an industrial spacecraft provider, and possibly one or more instrument providers. These institutions must commit to support the proposal effort itself as well as the subsequent implementation and science phases (NASA Phases A-E, concept development through operations, Box 2.1).[3] Large investments are required for competed space mission proposals of any size. These initial investments are used in proposal development, idea maturation, and trade space analysis. According to the Laboratory for Atmospheric and Space Physics (LASP) at the University of Colorado Boulder, and others an estimated $2 million to $3 million is typical for the development of a Discovery or New Frontiers mission proposal.[4] These commitments are not easily secured, both because of the up-front financial investment and the uncertainty of success inherent in the competition. Thus, decision makers at each institution find themselves in a situation where they must choose which PIs to support, if any; hence the "competition before the competition." This level of initial investment, before any return is even possible, represents a sizable barrier to potential PIs with differing impacts and ability to secure the resources.

**Finding:** The process of maturing a mission concept into a complete and feasible mission design and complete proposal requires large investments, often years before proposal submission, of time, money, and expertise. Access to the significant resources required for this process are not uniformly available to aspiring PIs.

Additionally, potential PIs must also be familiar with the organizations they may want to partner with and often have to navigate and adjust to a range of organizational cultures. There are no universal organizational processes for team formation and concept development; consequently, there are clear differences across institutions regarding access to mission design laboratories, flight dynamics resources, hardware and spacecraft bus design knowledge, proposal support, and scientific expertise that can disproportionately affect how teams are formed and who leads or even joins those teams. Respondents in the commissioned qualitative study of the experiences of mission proposers, cited networking as a primary means of proposal team formation, where factors such as institutional affiliation and personal reputation impacted how teams were assembled, as they shaped differential access to information,

[3] More on NASA mission phases can be found at NASA (2019).

[4] See more on mission costs in Table 2.1.

timelines, and processes. These potential inequities are not easily addressed, but they could be partially mitigated by improved transparency throughout the process. Procedural transparency in both the criteria for inclusion in the team, as well as in evaluation of potential team members, would allow for less reliance on networks and other social and knowledge capital, and better align decision-making about inclusion in the team with pre-specified rationale rather than post hoc justifications. Such transparency could include stated criteria of need and expertise requirements that are shared with potential team members prior to selection; making the assessment rubrics publicly available; and communicating directly with those ultimately not chosen about how to strengthen their chances in the future (Bielby 2000; Reskin 2003; Stainback et al. 2010). Chapter 5 discusses additional aspects of bias and the challenge of access to professional networks in more detail.

The greatest barriers can be at a single institution: who decides which individuals lead a particular mission concept? And whose ideas are financially supported? Through the testimony of several individuals in the space science community, the committee gained insight into the large variations that occurs surrounding these questions from institution to institution.

The committee heard from a range of organizations that serve as instrument builders, spacecraft providers, and research partners on NASA mission proposals. The types of organizations included for-profit businesses, universities, University Affiliated Research Centers, and nonprofit organizations that have led space missions. These representatives acknowledged that a small core group of "gatekeepers" are responsible for determining which proposal ideas move forward and are afforded access to the pool of B&P resources, contacts, support, and mentorship. In some cases, it is one individual who makes that decision for the institution. Other institutions described a process of internal gates wherein proposal ideas are evaluated by a proposal development committee composed of engineers, scientists, and management, and are assessed for likelihood of success before making institutional support decisions. The opportunity to use a committee approach like this exists at all major spaceflight proposing institutions, and selection processes need to continue to evolve based on the most recent evidence-based research on implicit bias and team dynamics. An even greater opportunity exists for these "internal" competitions to be open to "external" PIs and collaborators. For scientists who aspire to have a leadership role in a mission, activities such as involvement in mission data collection, participating in science team meetings, serving in review panels, and seeking mentorship from active mission leaders would help them learn firsthand about the mission process.

As discussed earlier, proposals may be submitted to a number of AOs before they are selected. The feedback from the review process, as well as the opportunity to mature technology can result in stronger proposals in later cycles. Critical feedback from experts outside of the team during the proposal development process is also a useful technique for gathering critiques and identifying potential proposal weaknesses before submission. Such outside expertise is more readily available through larger organizations and NASA centers, and lack of access to these experts can negatively impact proposal development and realizing a diverse pool of potential PIs. As previously mentioned, the institutional decision makers at some institutions and organizations become "zeroth-order gatekeepers" for missions and the PIs. Most of this power is concentrated in a relatively small number of institutions. Nearly two-thirds of the ~400 Step 1 proposals submitted over 20 years to SMD were managed by five institutions: Jet Propulsion Laboratory (JPL) (28%), Goddard Space Flight Center (GSFC) (22%), Johns Hopkins University Applied Physics Laboratory (JHU/APL), University of California, Berkeley (4.5%), and NASA Ames Research Center (ARC) (4.5%). Furthermore, of 93 proposals where a spacecraft provider could be determined, 78% of proposals included partnerships with only 6 organizations: Lockheed Martin (18%), Northrop Grumman (17%), Ball Aerospace (15%), JHU/APL (12%), Millennium Space Systems (10%), and Southwest Research Institute (SWRI) (6%) (Zurbuchen 2019).

Institutions often favor internal PIs such that the institutional investment remains "inhouse." Thus, PIs at smaller institutions with fewer resources may find it difficult to secure partners at larger centers for proposal development support. Figure 2.3 shows a preponderance of in-house PIs at the top mission management organizations, indicated by the proportions corresponding to "Self." While it is understandable that private and university-affiliated institutions would predominately or exclusively support in-house PIs, it is notable that 40-70% of the proposals supported by NASA centers had in-house PIs.

Figure 2.4 shows the submitting organizations for each of the divisions in SMD by discipline area between 1996 and 2019. Universities are the largest submitting organizations across all four disciplines, followed by NASA

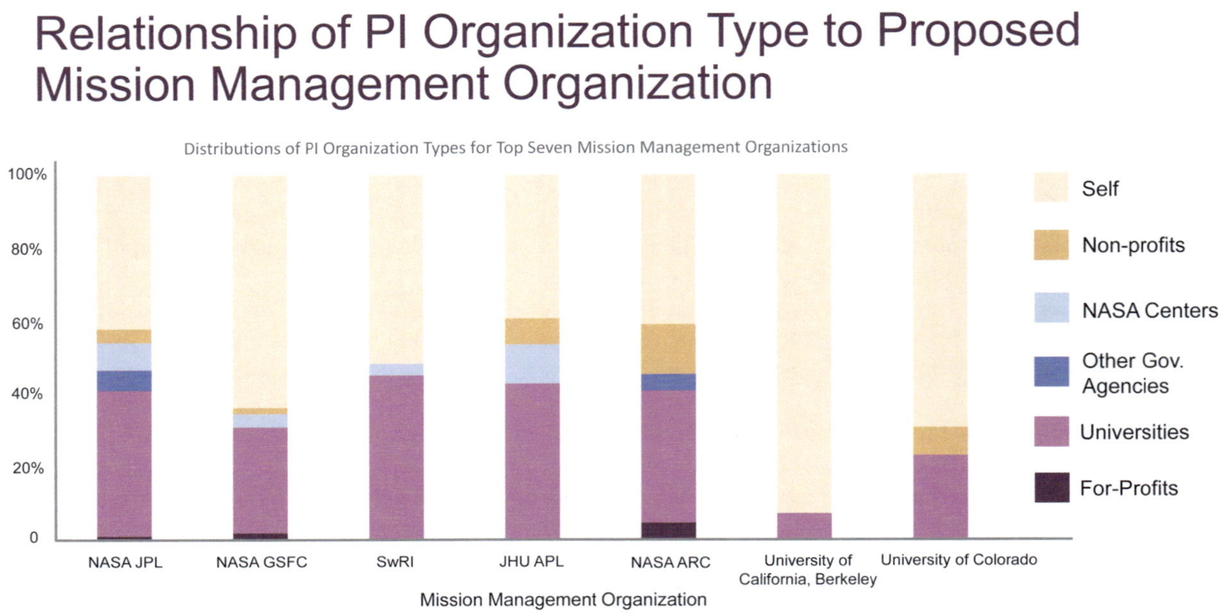


**FIGURE 2.3** Distribution of proposal PI organization to top mission management organizations from submitted mission proposals between 1996 and 2017. SOURCE: Zurbuchen (2019).

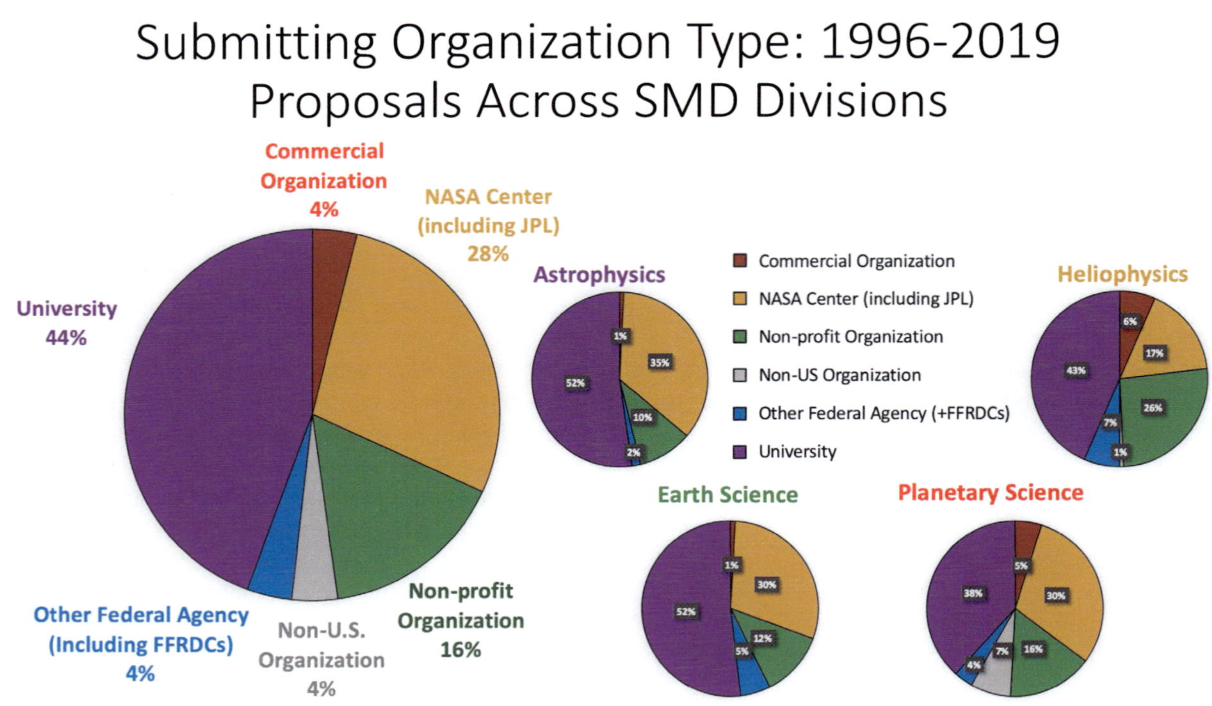


**FIGURE 2.4** Distribution of proposal submissions across NASA SMD by organization type. SOURCE: Heidi Jensen and Lorenzo Pappas, NASA Science Mission Directorate.

centers in all disciplines except Heliophysics. For Heliophysics, nonprofit organizations are the second-largest type of submitting organization. Thus, aspiring PIs may need to connect with leading universities, NASA centers, or nonprofit organizations for proposal development and submission.

**Finding:** Proposals to NASA mission calls involve collaboration of NASA institutions, spacecraft providers, and potentially instrument providers, research laboratories, and universities. Institutions invest hundreds of thousands to millions of dollars in developing concepts and assembling proposals. Decision makers at the institutions effectively control the investments and the opportunities to become a PI.

**Conclusion 2-1:** The process of team formation and concept development requires significant resources and is, in part, informal, idiosyncratic, organic, opaque, and often personality-driven. These informal and organic processes, influenced by who knows whom in the community, directly impacts the diversity of the PI candidate pool.

**Conclusion 2-2:** The variation of non-NASA/internal resources available for preparation and production of mission concepts, full proposals, and site visits may result in an unfair advantage of some institutions over others.

**Finding:** NASA centers supported ~28% of mission proposals. Of these, 40-70% had in-house (e.g., NASA) PIs.

**Conclusion 2-3:** NASA is a leading force in the overall demographic makeup of mission PIs. The process of PI selection for NASA center-led missions, however, could be much more transparent and committed to supporting traditionally underrepresented PIs. A restructuring of this type would provide a roadmap for other institutions to follow in their own PI selection processes.

## NASA PROCESS: PROPOSAL AND SELECTION MECHANISMS

There are five science divisions in NASA SMD: Heliophysics, Planetary Science, Astrophysics, Earth Science, and Biological and Physical Sciences. The competed opportunities and subject of this report focus on the first four divisions. Within these four NASA divisions, there are a number of programs for competed missions, with each program focused on a particular size/cost of mission. These are summarized in Table 2.1, which also includes the non-competed mission series for context: small to medium-class mission AOs (Explorers, Venture, Probes, SIMPLEx) are released approximately every 2 years, New Frontiers opportunities normally on a 3- to 4-year cadence, and flagships normally on a 4- or 5-year cadence.

The process for the release of AOs, response through proposal submission, and review of proposals has been improved and made more uniform across NASA's four divisions since 2009. As seen in Figure 2.5, the timeframe from the AO to the selection and initiation of a mission may take on the order of 1 year for one-step proposals or 2 years for two-step proposals (see Figure 2.6). These steps are almost entirely under NASA's direct control and strictly proceed according to procedures specified in federal regulations (the NASA FAR). Specifically, Section 1872 of the FAR governs the acquisition of investigations, with detailed specifications for the AO (see 1872.3), evaluation of proposals (see 1872.4; processes defined in 1872.401), and recommendation, selection, and debriefing (FAR section 1872.5). The mission directorate associate administrator (MDAA) retains the authority and responsibility for final approval of the AO and its release as well as for proposal evaluation and resulting selection(s).

**TABLE 2.1** Programs for Competed and Non-Competed Missions Across NASA SMD Science Divisions

| Heliophysics | Earth Science | Astrophysics | Planetary Science[a] |
|---|---|---|---|
| Flagship (not competed)<br>• Large, once-in-a-decade missions, >$1 billion<br>• Directed from HQ[b] to NASA centers<br><br>Other Directed Programs—costs vary<br>• Living With a Star<br>• Solar Terrestrial Probes<br><br>Other Competed Programs<br>• Instrument Opportunities<br>• Missions of opportunity (MoOs)/Rideshares: $35 million-$75 million (FY2019)<br><br>Explorer (competed)<br>• PI-led cost-capped: ~$165 million-$250 million<br>• Small-scale (SMEX): $165 million (FY2017)<br>• Medium-scale (MIDEX): $250 million (FY2019)<br>• MoO: $55 million (FY2019) | Flagship (not competed)<br>• Large, once-in-a-decade missions also known as "Designated Observables," >$1 billion<br>• Directed from HQ to NASA centers<br><br>Explorer (competed)<br>• PI-led cost-capped ~$350 million<br>• New Programs, details still in formulation<br><br>Earth Venture (competed)<br>• PI-led cost-capped:<br>• ~$35 million-$190 million<br>• Instrument (EV-I) hosted payload: $35 million-$108 million (FY2022)<br>• Mission (EV-M) free flyer: $190 million (FY2022)<br>• Continuity (EV-C) Demo of a continuity measurement: $150 (FY2019) | Flagship (not competed)<br>• Large, once-in-a-decade missions, >$1 billion<br>• Directed from HQ to NASA centers<br>• Instruments often competed<br><br>Probe<br>• New Program—details TBD: ~$750 million-$1 billion<br>• Concept studies funded in 2017<br><br>Explorer (competed)<br>• PI-led cost-capped: ~$145 million-$300 million<br>• Small-scale (SMEX): $145 million (FY2020)<br>• Medium-scale (MIDEX): $300 million (FY2022)<br>• MoO, SmallSat: $35-$80 million (FY2022) | Flagship (not competed)<br>• Large, once-in-a-decade missions, >$1 billion<br>• Directed from HQ to NASA centers<br>• Instruments often competed<br><br>New Frontiers (competed)<br>• PI-led cost-capped: $850 million (FY2015)<br><br>Discovery (competed)<br>• PI-led cost-capped: $500 million (FY2019)<br><br>SIMPLEx (competed)<br>• PI-led cost-capped: $55 million (FY2018)<br>• SmallSats |

[a] The SMD divisions also have suborbital programs (rockets and balloons).
[b] See Appendix A for definitions of acronyms.
SOURCE: Adapted and updated from S. Lipsy, "PI-Led Mission Proposal Process," presentation to the committee, February 8, 2021.

### NASA Proposal Process Overview

The overall flow of the NASA proposal process starting with the AO using a one-step process is shown in Figure 2.5. For some larger missions, a two-step process is utilized, as illustrated in Figure 2.6. The one-step process and the first step of a two-step process are identical. Proposals are evaluated against a set of criteria around science and technical factors that are discussed in more detail in later sections. For a one-step proposal, the evaluation reviews flow through the categorization subcommittee step below where one committee integrates the evaluations into one of four categories ranging from recommended for acceptance to recommended for rejection (details in later section) and the steering committee then recommends selections. The decision-making authority, the SMD associate administrator (AA), then finalizes the selection of missions to go into formulation. For two-step proposals, the evaluation feeds into Categorization and Selection for the Concept and Technology Development Phase (Phase A). All missions selected then complete a CSR on the same schedule and a review step (technical, management, and cost [TMC] feasibility required, science evaluation if necessary) is used to down-select to a final mission selection for formulation. In the sections that follow, the main steps of this process will be discussed in more detail.

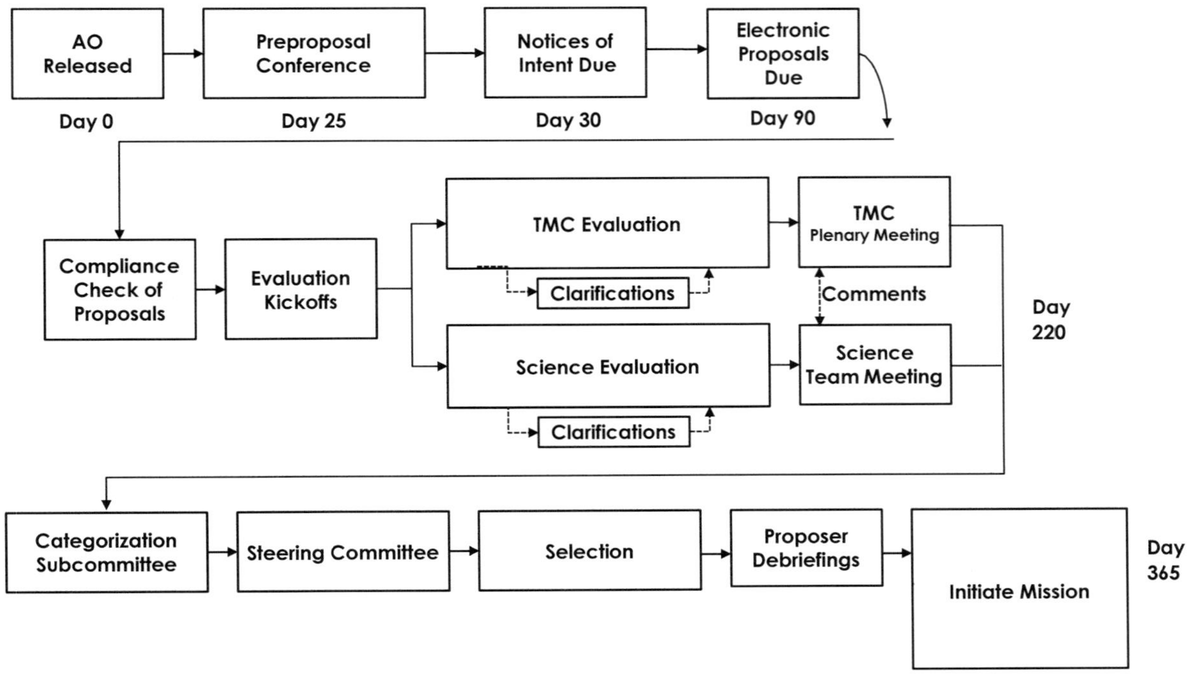


**FIGURE 2.5** Typical process flow for one-step process of NASA competed missions and instruments with notional timing. SOURCE: Thomas Wagner, NASA Science Mission Directorate.

## NASA AOs and Proposal Submission

One unique aspect of the NASA AOs is the FAR Supplement that governs the processes. NASA is one of the few federal agencies with a supplement of this nature because of the unique nature of its work with the science and technology communities and the sizeable investments it makes to obtain science data derived from investigations to, from, and in space. An important element in the FAR is Part 1872: Acquisition of Investigations, under which AOs for competed opportunities fall. The regulations spell out in rather specific detail the process for solicitation, review, evaluation, and selection in addition to roles, accountability, and authority.

As described by the FAR, the AO system used for acquisition of science, technology, or exploration flight investigations is different and separate from the agency procedures for acquisitions for prior known requirements. A general distinction between normal acquisitions for goods or services and AO investigation acquisitions is that AO proposers define not only how they will accomplish investigation objectives but also the objectives themselves within the framework of broader articulated NASA goals. Unlike a proposal responding to an RFP, an AO proposal effectively defines a statement of work in addition to the approach (and cost) for how the proposer plans to carry it out. Typical examples where the use of AOs is most applicable are solicitation of complete small- and intermediate-size investigations (larger and more costly than suborbital-class research but smaller and less expensive than the largest and most expensive strategic flagship missions), of instrument investigations on larger missions whose overall management is assigned to a NASA center, or of spaceflight instrument investigations contributed to a non-NASA mission.

The Science Office for Mission Assessments (SOMA) at NASA Langley Research Center (LaRC) supports SMD at NASA Headquarters in the acquisition of Earth and space science missions and instruments through the development of AO solicitations and the TMC evaluations of proposals received in response to the AO solicitations

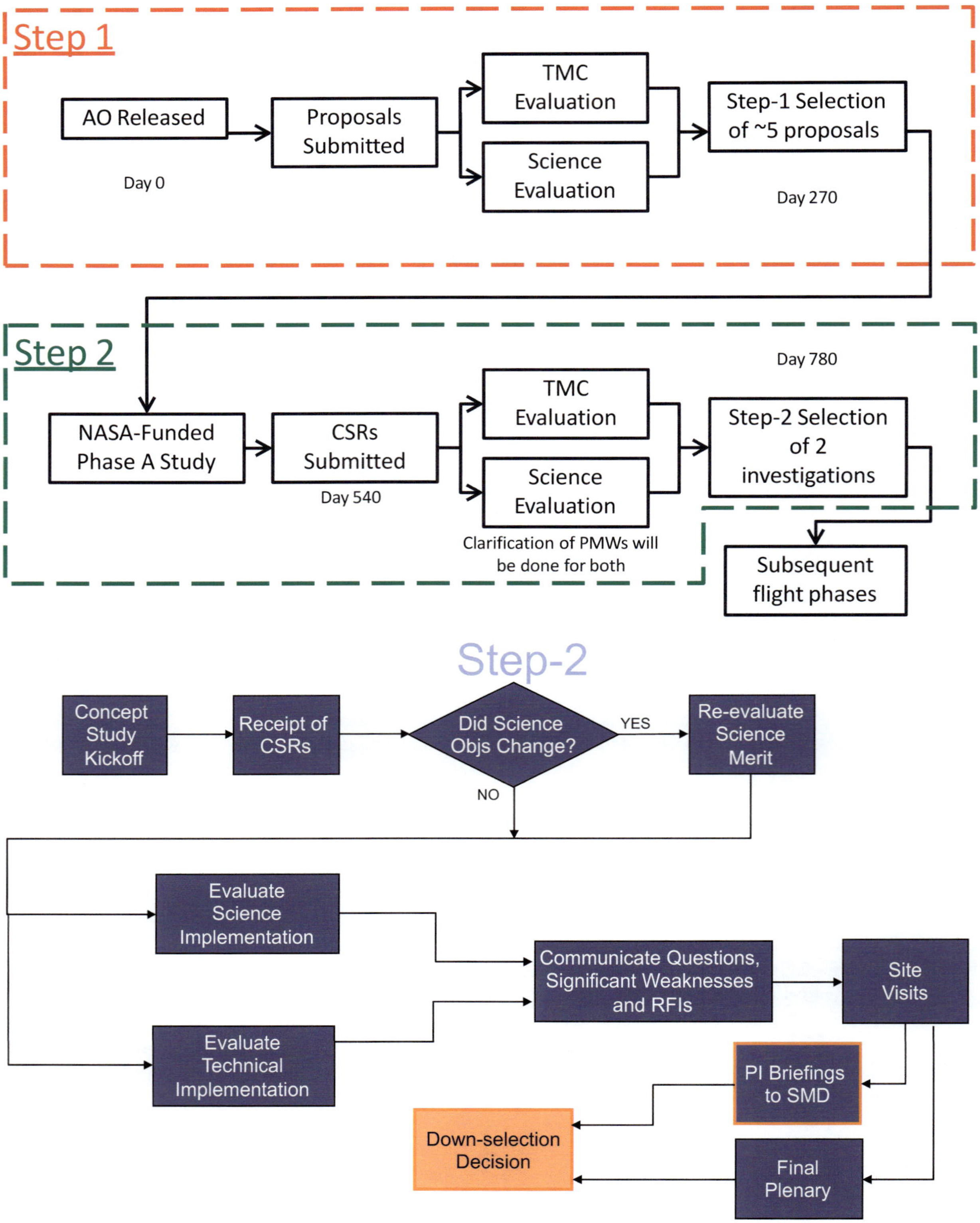


**FIGURE 2.6** (*Top*) The typical two-step proposal process at NASA with a notional timeline. CSR here is the Concept Study Review, discussed in a later section. NOTE: PMW, potential major weakness. (*Bottom*) The second step in a two-step proposal process at NASA in more detail. SOURCE: (*Top*) Thomas Wagner, NASA Science Mission Directorate. (*Bottom*) Hamden et al. (2020). See https://science.nasa.gov/researchers/pi-launchpad.

and Phase A concept studies.[5] This office serves as a resource center with a library of AOs, lessons learned documents, and the schedule of upcoming AOs.

NASA has created a centralized portal for information about proposal opportunities and the submission of proposals called the NASA Solicitation and Proposal Integrated Review and Evaluation System, or NSPIRES. Proposals of all kinds are submitted through the NSPIRES system, and all submissions are electronic. In the case of mission proposals, a typical proposal length is 200-300 pages, with 20-30 pages focused on science. Given the complexity of the evaluation criteria (to be discussed below), a proposal must be extremely well crafted to respond to all of the criteria. Preparation of such a large document, including proper formatting of figures and cross-referencing of sections and information is a significant effort. In addition, NASA flight missions must comply with NASA Space Flight Program and Project Management Requirements, known as NASA Procedural Requirement (NPR) 7120.5 (NASA 2021), which defines the NASA life cycles for spaceflight programs and projects, as well as program and project management roles and responsibilities.

**Finding:** The NASA AO process is highly structured and bound by FAR and the NASA FAR Supplement, which fully articulate it; however, in practice, the process is complex and can be very difficult to navigate and understand.

### NASA Evaluation and Selection Processes

As described in the FAR and illustrated in Figures 2.5 and 2.6, once proposals are submitted, the NASA process begins with evaluation. The evaluation results feed into categorization of the proposals, and then the Steering Committee validates the categorization. Recommendations for selection are made based on the categorizations, discussed further below, but are influenced by the division programmatic considerations. The division prepares a briefing for the selection official, who may, and typically does, consult other agency officials, but has the sole authority for the selection.

The scientific and technical evaluation criteria, at the highest level, are quite streamlined. The three high-level criteria are: (Criteria A) scientific merit of the proposed investigation (4 factors); (Criteria B) scientific implementation merit and feasibility of the proposed investigation (5 factors); and (Criteria C) TMC feasibility of the proposed mission implementation (5 factors). These are described in detail in the FAR in Section 1872.402, as follows Criteria and Factors (see Annex 2.B, Tables 2.B.1-2.B.3 for descriptions of the factors). These are the evaluation factors applied in one-step proposals and in the first step of two-step proposals. The weight of the evaluation factors must be stated in the AO if they are unequal. The typical weighing is science merit (Form A) 40%, science implementation (Form B) 30%, and TMC (Form C) 30%. Annex 2.B, Tables 2.B.1-2.B.3 reveal the complexity of the evaluation factors, which contributes to the large level of effort and high costs of producing a competitive proposal.

The evaluation criteria do not specifically include diversity, which is a missed opportunity, as the evaluation criteria drive the proposal development strategy and the focus of efforts and resources. Factor B-5 (probability of science team success) does include the probability of the investigation team success, including the experience, expertise, and organizational structure of the science team. In September 2021, NASA released a request for information (RFI) with the title "Adding Inclusion, Diversity, Equity, and Accessibility (IDEA) Requirements to NASA Announcements of Opportunity." One aspect of the RFI was to solicit feedback on the proposal to add this evaluation criteria "Factor B.6. Inclusion, diversity, equity, and accessibility (IDEA). The team's IDEA plans for forming a diverse team, and plans for creating and maintaining an inclusive and equitable environment will be assessed. This factor includes the alignment of the proposal with NASA's core value of inclusion and the likelihood of successfully achieving the objectives of the "Code of Conduct" in service of mission success (NASA 2021). The final status of this proposed criteria is unknown at the time of the writing of this report, but if adopted, this is a positive step in including an aspect of diversity, equity, inclusion, and accessibility (DEIA) in the evaluation of proposals.

[5] Standard template of the AO can be found in NASA (2018).

**Finding:** Team diversity is not included in the criteria used to evaluate NASA mission proposals, but currently may be included with "other factors" for consideration by the selection official in final selections. A NASA RFI was released in September 2021 seeking comment from the public on the proposed addition of an evaluation criteria focused on the DEIA plans of the proposing teams.

The evaluation of Criteria A and Criteria B are performed by panel(s) of community experts who are free from personal and organizational conflicts of interest. Criteria C are evaluated by a TMC peer review panel. For each Factor in each criterion, the panels develop assessment findings or state that there are none. The panel chair prepares a report form on findings that captures the major points, and these are expressed in short narrative for identified major strengths, minor strengths, major weaknesses, and minor weaknesses. A summary evaluation is then developed, ranging from Excellent to Poor for Forms A and B, and from Low Risk to High Risk for Form C. The rubric for these summary evaluations is included in Annex 2.C where Table 2.C.1 contains details for the science and implementation evaluation, and Table A.5 tabulates the TMC summary evaluation details.

The results of the proposal evaluation Form A, Form B, and Form C are then used by the Categorization Committee and Steering Committee in the process described in FAR Section 1872.404. These committees are composed wholly of civil servants and Intergovernmental Personnel Act appointees. The Categorization Committee will consider the evaluation results and categorize the proposals as defined in the AO, in levels from Category I (recommended for acceptance) to Category IV (recommended for rejection). The details of all four categories are presented in Annex 2.D, Table 2.D.1. The categorization process integrates the information and scoring from the evaluation against Criteria A, B, and C, and the written assessments. As stated in the FAR, the objective of the categorization process is to derive a portfolio of consolidated merit-risk characterizations that will be subjected during the recommendation and selection processes to additional programmatic criteria. In this way, all of the science and technical evaluation information is considered by one body and categorized by one committee. The inclusion of programmatic criteria is important for NASA to achieve its broader goals, but may appear to proposers as a lack of transparency.

The Steering Committee then reviews the results of the proposal evaluations and categorizations, and conducts an independent check on the quality, balance, and integrity of the evaluation process to this point. This committee requires less detailed technical expertise and can provide a fresh perspective from qualified and experienced individuals with broad or different backgrounds. Details of the steering committee responsibilities and expectations are in FAR Section 1872.405.

In the case of one-step proposals, the results of the proposal evaluation will be presented to the SMD AA, who will make the final selections. The overriding consideration for selection will be to maximize scientific return and minimize implementation risk while advancing NASA's science goals and objectives within the available budget for this program. In addition, the SMD AA may take into account a wide range of programmatic factors in deciding whether or not to select any proposals and in selecting among top-rated proposals. Programmatic factors can be administration priorities, recent discoveries, balancing portfolio, etc. In this way, NASA selection officials may also act as gatekeepers.

In the case of two-step proposals, multiple missions are selected and funded by NASA at this step to begin the Concept and Technology Development Phase (Phase A). The product of the Phase A study is the CSR, which corresponds to a CML of 6. The CSR allows the TMC to see additional detail, a higher fidelity cost estimate, and schedule consistent with expectations for Phase A products. The science objectives and science requirements are re-validated or amended if needed. If the science, exploration, technology objectives, or mission requirements change, Form A (the proposal's intrinsic merit) will be re-evaluated. In all cases, Form B and Form C are reassessed by a Concept Study Review Panel. The requirements for the CSR are delineated in a 50-page document (NASA 2010), and example slideshow presentations that outline the CSR evaluation plan are publicly available.[6] The overall flow of evaluation steps is shown in Figure 2.7, and the presentation of the CSR is conducted in a full day in-person meeting or in a virtual format, where 10 hours of presentation occurs over 2 days. Site visit

[6] See, for example, NASA (2021a).

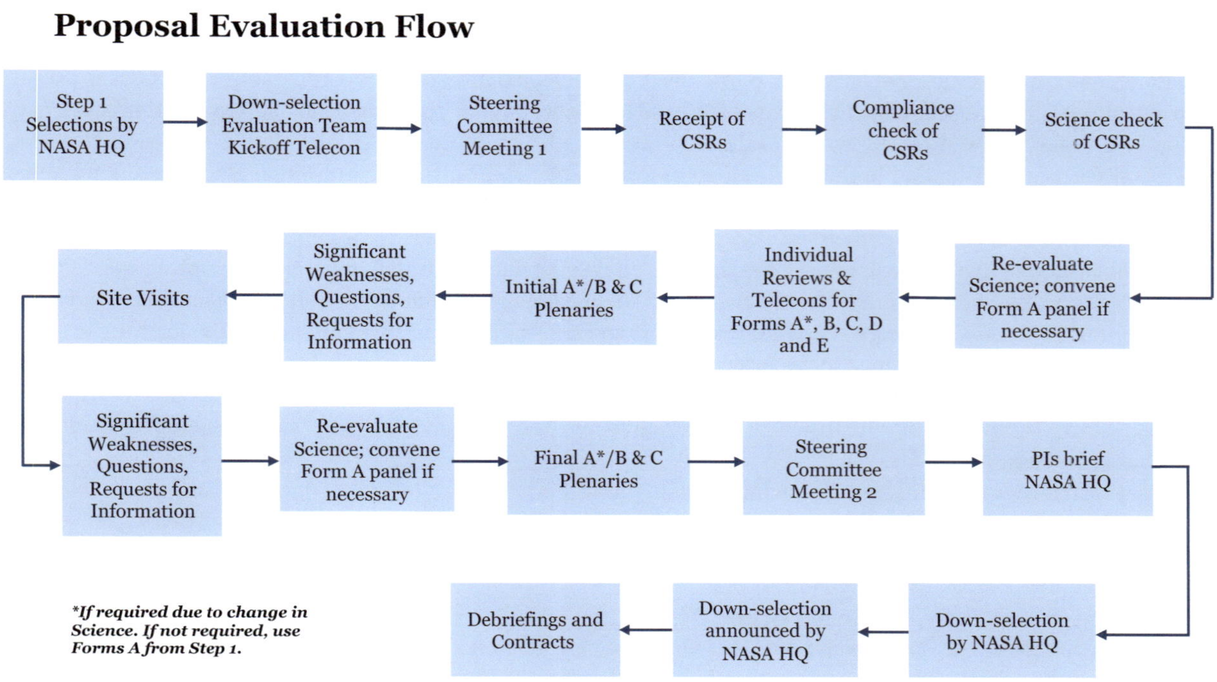


**FIGURE 2.7** Overall flow of the evaluation steps for Concept Study Reports. SOURCE: NASA Science Mission Directorate. See NASA, 2021, "2019 Heliophysics Explorers Program (HP) Medium-Class Explorer (MIDEX) Concept Study Report (CSR) Evaluation Plan," Announcement of Opportunity NNH19ZDA0130, https://explorers.larc.nasa.gov/HPMIDEX/pdf_files/HPMIDEX19_DS_Eval_Plan_20210804.pdf, Slide 8.

instructions are provided to teams with details of the actions, timing of events, and tools to be used for site visits. The PIs brief NASA Headquarters in the final stages of the CSR evaluation.

NASA provides some funding for Step 2 proposals with the expectation that all Phase A and products requirements will be developed and ready for in-depth review, consistent with the CSR requirements including independent evaluation by a group of experts during a site visit. Typically, at this stage, many institutions, including NASA centers and some universities, provide additional funds to supplement the NASA Phase A funding. The results of the evaluation of the CSR and site visit are used to select a mission for continuation (down-select) and recommendation to the selection official. The process is rigorous and, in many cases, taxing, and the NASA-provided funding can be inadequate to meet expectations for the CSR and the preparations and execution of the site visit. Moreover, limitations in resources beyond the NASA provided funds can specifically put smaller colleges, universities, companies, and nonprofits at a disadvantage as well as new PIs who are not aware of this additional need.

> **Conclusion 2-4:** Preparation for both the CSR and the associated site visit for two-step proposals are intense and demanding activities, involving 1 to 2 days of the site visit and a subsequent presentation from the PI to the SMD AA. Some institutions provide additional funds to supplement the NASA Phase A funding. Thus, the variability of resources available and the critical role of a single presentation (PI to SMD AA) merit reconsideration of the resource allocation, review content, and purpose.

Competed mission opportunities vary in scope, complexity, and cost boundaries. What is required to put together a winning proposal (i.e., compelling science investigation, technically feasible, and programmatically credible) varies accordingly. It is a challenging job with great expectations, particularly for the PI who by definition has the most at stake in the process in terms of time investment and expectations for accountability for the scientific/technical and programmatic success or failure of the mission.

### Feedback to the PI Teams

PI teams expect and deserve feedback, particularly non-selected investigators. Constructive feedback is useful, even when there are critical and major weaknesses. The highest value component of the feedback process may be the face to face debrief during which evaluation process leaders do their best to explain weaknesses that may have been most discriminating in the final selection process by senior officials. Anecdotal evidence suggests that feedback to non-selected PI teams on mission proposals is uneven, particularly for Step 2 proposers who are not selected, because decisions are sometimes driven by other factors and are more programmatic in nature. While there may be other factors that influence the final selection, conveying such information would be helpful especially for mission concepts that fall in the selectable categories. The feedback will shape reinvestment strategies to bid again armed with helpful tools and feedback. The feedback is an essential element and a big part of the learning process, particularly for new investigators. Going forward in the highly competitive mission AO competitions, attention to the quality and constructive utility of the personal and written feedback is almost as important as debriefing the selected teams. The attention will foster an encouraging environment and assist to better develop a pipeline of high quality mission proposals.

The process for evaluating and reviewing mission proposals and the information presented and made available by NASA have become more uniform across divisions, with improvements in transparency and clarity. Nevertheless, it is a complex and long process with detailed, extensive evaluation criteria that can be daunting for those going through the process for the first time.

**Conclusion 2-5:** The steps of the NASA AO process are governed by detailed regulations, but scrutiny for sources of bias throughout the implementation of the process is warranted, and the process would benefit from including DEIA as criteria for evaluation.

## PI AND TEAM LEADERSHIP DEVELOPMENT: CHALLENGES, OPPORTUNITIES, AND POTENTIAL BARRIERS

As has been described, all aspiring PIs must overcome numerous challenges stretching over a period of several years or more. These impediments are likely to be exacerbated for historically underrepresented populations, including women, or for those from research institutions that lack a record of NASA missions. Challenges discussed earlier in this chapter include:

- Gaining mission leadership experience
- Benefitting from mentorship
- Winning commitment from one's home institution
- Forming a suitably broad and deep team,
- Establishing leadership
- Climbing the CML and TRL levels
- Securing institutional commitments and B&P resources (competition before the competition)
- Leading preparation of a compelling proposal
- Leading preparation of a compelling CSR
- Conducting an appropriate site visit
- Briefing the SMD AA

In the final section of this chapter, additional challenges are discussed, including but not limited to team dynamics, access to information and training, and proposal review.

## Team Dynamics

A wide range of expertise is brought together for a mission proposal. The PI has the ultimate responsibility carrying out and reporting the results of the mission (NASA 2021d), but there are many other key roles. For example, the project manager is responsible for the formulation and implementation of the mission in accordance to NPR 7120.5, and the engineering technical authority or system engineer manages the engineering activities, including systems engineering, design, development, sustaining engineering, and operations. A deputy PI may be on the team, sharing responsibilities with the PI. An instrument lead may be included to provide engineering leadership for the instrument. For PI-led competed projects, there may be a project scientist, who is part of the PI team and works closely with the PI. The project scientist is typically delegated the responsibility to monitor the scientific output of the project and ensure that the project achieves each of its science requirements. For directed projects, there is no mission PI. The project scientist in that case is responsible for a more significant fraction of the project-level management than in a competed project. The project scientist works closely with the project manager and is directly responsible for all science related tasks. Other key roles are the operations manager, who oversees the planning of flight operations and instrument commanding. A ground data systems lead manages the data processing system. The PI is critical for the mission; however, the overall success and team dynamics are driven many contributors that require technical expertise and leadership skills.

There is no defined process by which teams form in the early development space between CML 1 and 2 or 3. To this end, the committee relied heavily on the scholarship on team science, interdisciplinarity, and team dynamics to make determinations about how the undefined nature of team formation for missions impacts diversity in the proposal pool. In their presentations to the committee, scholars who specifically study NASA spacecraft teams as well as team effectiveness in science described team formation as "organic," meaning that it is tied to who you know and who you are mentored by. For example, in working with NASA robotic spacecraft teams, Vertesi (2021) found that teams form across NASA institutions and universities in a "federation of cultures," and with precious, few missions and interplanetary timelines, teams work together for multiple decades on the same project, and lifetime appointments in the same positions do not allow for organizational or intergenerational mobility. Network effects and relationships also matter, as PI-led missions reduce costs up to 30% by relying on existing ties (NRC 2006). Other scholars reported how "high-performing teams" tend to stay together from project to project, which may discourage new researchers. Also, minorities or other marginalized groups may be subject to tokenism (Kanter 1977b) or backlash (Rudman and Glick 2001), leading to continued low representation on teams (Correll 2004; Rathbun et al. 2018; Porter et al. 2020; Rivera-Valentin et al. 2020). An opportunity exists to integrate emerging researchers as mentees and participants across the mission teams, including the higher leadership positions. In practice, this may be through formal mentorship, training of existing team members to prepare them for advancements, and strategies to expand participation of emerging researchers in science team meetings. Such activities present enormous opportunities for emerging, early career researchers to rise through the ranks, gaining invaluable experience in running a team and mission along the way.

The scholarship on the science of team science highlights that team culture, team dynamics, collaboration readiness, decision-making strategies, intersectional identities, cultural biases toward women and underrepresented groups in leadership positions, and inter-organizational cultural differences are all important factors that influence who is on the team and how well the team will perform. For instance, working across institutions was found to present challenges to teams as inter-institutional collaboration may involve variable chains of commands and intercultural communication challenges akin to international teams (Ting-Toomey and Oetzel 2001; Lebaron and Pillay 2006; Hall et al. 2008; Vertesi 2020). In a study by Rozovsky et al. (2012) on how teams work, it was found that psychological safety, which allows for "moderate risk-taking, speaking your mind, creativity, and sticking your neck out without fear of having it cut off" was the most critical factor for the success of high-performing teams. Additionally, using an intersectionality framework allows for a better understanding of how social issues restrict access and power, and without this framework one defaults to listening to only the most privileged groups (Cole 2021). Chapter 5 of this report also discusses team dynamics and insights of the impacts on PI-led missions. Increased awareness of the evidence-based considerations for science team effectiveness by all who participate in NASA-funded missions can influence future team formation and team performance. Efforts to form teams and bring together the needed skills and knowledge would benefit from considering current research on key topics, such as how the most excellent team science gets done, what are the impacts of cultures on diverse teams, and how to use consensus-based decision-making. NASA can encourage these concepts to be integrated at the start of team formation by providing best practices documents and training sessions that educate PIs and team members on such strategies as the Concordance Method of decision-making (Schutz 1994).

The expectation of what makes a "good PI" may be outdated and misdirected. For example, a National Academies study from 2015 states "leadership is not a quality that an individual either has or lacks, and there is not a single leadership style that is effective in all contexts (NRC 2015). Yet, there is a myth that persists in the fields that PIs must have certain personality traits such as dominance, paternalism, sole decision-maker, and more. It is also communicated to potential PIs that "your life will be taken over, including evenings and holidays" (Zurbuchen 2019). However, that is not a prerequisite. While the role of a PI is significant and a large undertaking, leading a space mission is a team endeavor. The leadership of a large technical team and execution of a complex technical mission is demanding, but the leadership of such teams can be executed effectively in many different ways, as described in a recent National Academies report. Discussions of the expectations of PIs by NASA do not currently reflect the wide range of leadership styles.

**Finding:** A frequently heard message is that being a PI is all-consuming and nights and weekends will need to be sacrificed to take on this role. Other data show that different PIs have different styles of management and delegation, and there are PIs who effectively employ delegation, time management, and boundary setting.

**Conclusion 2-6:** NASA can do more to socialize the concept of diverse leadership styles and highlight that leadership success is more important than leadership style.

### Information Access and Training

Each AO has attracted a reasonable number of PI-led proposals that achieve the highest ranking and are eventually selected and flown (see section "NASA Process: Proposal and Selection Mechanisms" of this chapter for more discussion of selection process). These proposals have originated in NASA centers, university affiliated research centers, and in a select set of universities and public or private research centers that either have the capability of carrying out such large-scale space projects or can collaborate with these institutions. Within the past few years, NASA has recognized that such an organic approach limits the population of potential PIs and can limit access to the pool of skills and ideas that would come from broader participation. Findings from the commissioned qualitative study (see more in Chapter 5; also see Appendix C) reinforce the perspective that early career mission experience, mentoring, and access to networks of colleagues across a range of institutions are seen as very important for developing PIs, yet can be difficult to access. The commissioned qualitative study also revealed that respondents felt that the dedication and qualifications of the proposed PI play an important role in mission success; however, all of the skills needed are not learned through formal education routes like undergraduate and graduate training. Some respondents also believed that NASA training opportunities are not always widely publicized and are more accessible to those who are well-connected. These topics are also further discussed in Chapter 4 and Chapter 5.

The PI Launchpad initiative is one mechanism that was motivated by an aspiring PI who was frustrated with the process of proposal development. Created in collaboration with NASA and the Heising-Simons Foundation in 2019, this still evolving training is currently a 3-day workshop targeted at researchers and engineers who would like to submit a NASA space mission proposal in the next few years but do not know where to start. The workshop is designed to introduce the proposal process and some of the key steps, such as creating a science case, developing requirements, writing a science traceability matrix, building a team, and getting support at your home institution.[7] Overall, this initiative exists to better prepare and aid the scientific community with the proposal preparation and submittal process.

[7] A complete list of areas covered by the PI Launchpad Initiative can be found in the workbook NASA (2019).

In addition to the PI Launchpad, other resources for aspiring mission PIs are offered by NASA.[8] These include the following: a recording and slides from a colloquium presentation by Thomas Zurbuchen, AA of NASA SMD, titled "Writing Successful Proposals: Observations from NASA" (Zurbuchen 2019); recordings and slides from presentations on writing Research Opportunities in Space and Earth Sciences (ROSES) proposals; proposal writing workshops at the American Geophysical Union and other conferences; planning resources for AOs, such as the standard template used; and mission management resources such as NASA's procedural requirements. Additionally, a general guidebook for proposers is routinely updated by NASA (2021b). Aside from this resource, few professional development offerings by NASA exist for prospective mission PIs, thus requiring a reliance on other information from the field to navigate the proposal process. A good example of this is a host of resources put together by a PI of an Earth Science Mission proposal (Gentemann 2021).

As discussed earlier in this chapter, a significant commitment of time and money is needed to develop concepts into winning proposals. In addition, the process can be intimidating to new proposers, particularly those at the early career stage, and those that historically have not had the benefit of a support system that values diversity and inclusion. This suggests that unless there is significant support, awareness of the challenges, and sustained mentorship in place it can be hard to start and remain committed to the process while navigating the barriers that exist along the way, particularly for women and underrepresented investigators. Investigators with a broad network of support either within their own or a partner institution or through other means of exposure and interactions are obviously at an advantage. A key question is whether the training opportunities to be exposed to the process and support systems are available to others in an equitable way. This is a complex issue that requires a multi-pronged approach and investment in capital if realizing DEIA in the PI proposer pool is a goal.

A variety of issues are at play, many of which may not be within NASA's control. For example, are the intersectional identities of proposers taken into account or are they defined by only one demographic to which they belong? Also, what is the current state of the space sciences pipeline and are national level STEM-related programs and priorities, such as NSF ADVANCE (Organizational Change for Gender Equity in STEM Academic Professions) and NASA Science Activation, having the intended effect and in a timescale that allows participants to see themselves growing up the ladder into positions of leadership? Looking beyond the pipeline, it is important to understand the venues and opportunities whereby meaningful contributions (i.e., not just token participation) with consideration for the full expression of DEIA (e.g., a PI, deputy PI, investigation team lead) is enhanced and nurtured. Figure 2.8 shows the distribution of science team members across NASA mission proposals from 2006 to 2020. There are more Co-Is than any other role (81%), and PIs are only 5% of the pool. Interestingly, there are less than 1% of deputy PI roles, suggesting very few PIs are planning on working with a deputy. Training and opportunities to gain mission leadership experience exists in these other roles, especially the Co-I role, as well as the deputy PI role.

**Finding:** Aspiring PIs are expected to have some experience on flight missions or instrument development, and will need networking and management skills. This type of experience and skill development is not typically acquired during undergraduate or graduate degree work. There are some opportunities in existing missions, but they are extremely limited. Many aspiring PIs reported frustration with the limited opportunities for training and gaining relevant experience.

**Conclusion 2-7:** Additional opportunities for training and mission experience are important for expanding the group of trained professionals that become potential mission PIs, and could range from formal mission roles on mission teams; to mentorship from current and previous mission leaders; to resources and strategies for increasing the involvement of Earth and space science community members in mission science team meetings.

[8] NASA, "New Principal Investigator (PI) Resources," https://science.nasa.gov/researchers/new-pi-resources.

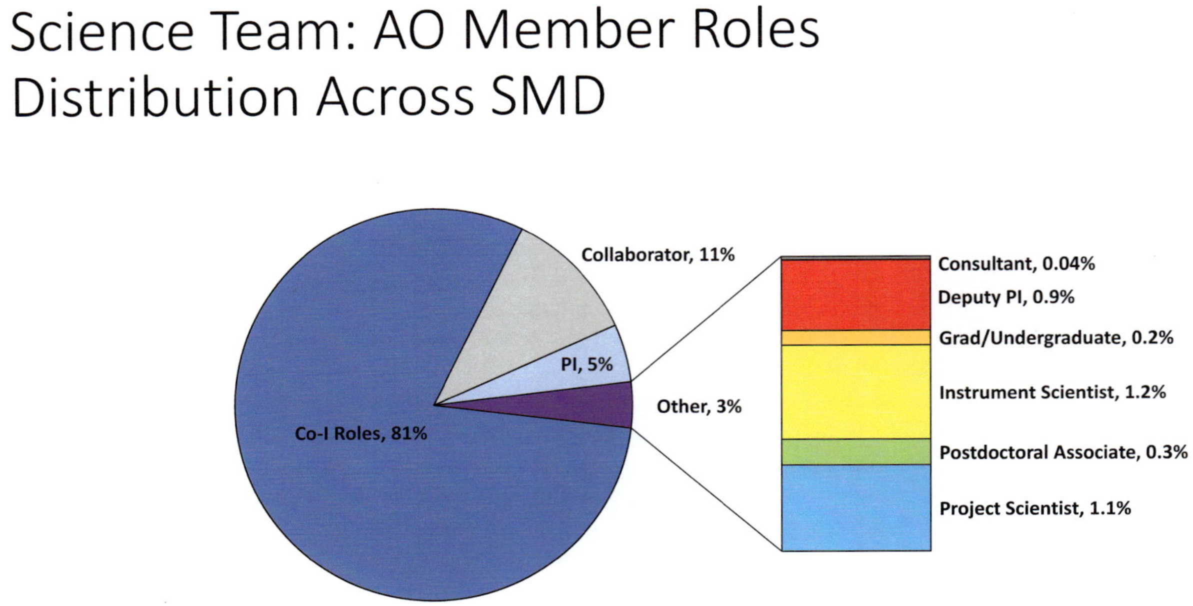


**FIGURE 2.8** Distribution of roles of science team members on SMD proposals from 2006-2019. NOTE: Science team data is only available 2006-2019. SOURCE: Heidi Jensen and Lorenzo Pappas, NASA Science Mission Directorate.

### Proposal Review

Proposal review is a process where bias can impact the results. NASA seeks to remove barriers to participation and reduce the impact of bias in the process, and one approach to achieving this objective is to ensure that the review of proposals is performed in an equitable and fair manner. Recently, NASA has explored Dual Anonymous Peer Review (DAPR) processes in its ROSES calls following the successful implementation at the Space Telescope Science Institute (Strolger and Natarajan 2019). Other similar studies conducted at the National Radio Astronomy Observatory, the Atacama Large Millimeter Array, and the European Southern Observatory have also indicated biases affecting proposals (Lonsdale et al. 2016; Patat 2016).

NASA has begun pilot implementations of DAPR in ROSES-2020 and early data appear consistent with improvements, both in terms of the overall quality of the review process, as well as in the demographics of awardees. For instance, in the Astrophysics Data Analysis program, prior to dual-anonymous review, women constituted 26% of the applicant pool, but only finished in the top two places in the panels' rankings 16% of the time. Following the switch to dual-anonymous peer review, women constituted 31% of the applicant pool and finished in the top two places 32% of the time.[9] Moreover, the success rate of early-career investigators even eclipsed that of the more senior investigators, further enriching the talent pool. The evidence to date continues to underscore the value of removing any possible sources of bias, while recognizing that it is not a substitute for other steps to improve the diversity of the pool of applicants for all NASA competed opportunities. The DAPR process and principles can be a vehicle to also improve the diversity of competed mission investigator teams and deserve attention and methodical experimentation by NASA throughout the proposal evaluation and selection process.

**Finding:** NASA SMD has begun to test and implement changes to peer review to broaden applicant pools and reduce potential biases in evaluations and awards. The observed changes after instituting DAPR were a broadened applicant pool and selections with gender ratios closer to submission gender ratios.

[9] NASA, "Overview of the Dual-Anonymous Peer Review Process," https://science.nasa.gov/researchers/dual-anonymous-peer-review.

## ANNEX 2.A

### Concept Maturity Levels

The committee adapted the following concept maturity levels (CMLs) from Wessen et al. (2013) to provide more complete details and typical resource requirements to achieve each level.

- *CML 1 Sketch on a Napkin*—The science questions have been well articulated, the type of science observations needed for addressing these questions have been proposed, and a rudimentary sketch of the mission concept and high-level objectives have been created. The essence of what makes the idea unique and meaningful have been captured. The conceiver of the idea may not be the principal investigator (PI) (e.g., an orbital dynamicist may identify an interesting trajectory that may enable unique science, and approach a scientist who has a potential to lead the development as the PI). CML 1 idea is shared with a person/entity with capabilities/resources to test initial feasibility to advance to CML 2.
- *CML 2 Initial Feasibility*—The idea is expanded and questioned on the basis of feasibility, from a science, technical, and programmatic viewpoint. Lower-level objectives have been specified, key performance parameters quantified and basic calculations have been performed. These calculations, to first-order, determine the viability of the concept. Initial feasibility study is usually a low-level effort that may or may not be funded. Whether to invest time, effort, and energy into this activity is often a personal or career judgment on risk, benefit, ambition, and interest. When funding is available, it is funded internally (e.g., Internal Research and Development [IRAD] programs). Technology maturation needs may be identified from this work.
- *CML 3 Trade Space*—Exploration has been done around the science objectives and architectural trades between the spacecraft system, ground system and mission design to explore impacts on and understand the relationship between science return, cost, and risk. Proposing team makes contact with a mission design center to perform a trade study (CML 3). Variations exist in the manner of contact—mission proposal team may contact a design center, or vice versa. In other cases, a mission concept may organically arise between the proposing team and the design center. Design center performs a design trade study to achieve CML 3. A trade study is a high-level effort, typically involving a small science team (~5 members) and 10-20 engineers and takes place over a 2-3 month period. Technology maturity assessments are likely to be an aspect of the trade study. Proposing team may be affiliated with the institutions that has a design center capability, in which case the process is fully internal to the institution. When the proposing team/PI are not affiliated with the design center's institution, the effort of the PI/Proposing Team may not be funded, or a contract may be involved between the design center and the proposing team. In some cases, the proposing team pays the design center, while in others the design center may fund the proposing team.
- *CML 4 Point Design*—A specific design and cost that returns the desired science has been selected within the trade space and defined down to the level of major subsystems with acceptable margins and reserves. Subsystems trades have been performed, and technology readiness level assessment is further refined. Design center builds a point design to advance the concept to CML 4. A point design process is a high-level effort by the science (5-10 members) and engineering teams (~20 members, usually the same members as the trade study team). Sometimes, CML 3 and CML 4 efforts are decided together as the continuity in the personnel and effort is beneficial.
- *CML 5 Baseline Concept*—Implementation approach has been defined including partners, contracting mode, integration and test approach, cost and schedule. This maturity level represents the level needed to write a NASA Step 1 proposal (for competed projects) or hold a Mission Concept Review (for assigned projects). Design center builds a baseline concept to be included in a proposal. Some institutions require Blue/Red team reviews at the design center to down-select concepts within the institution to avoid internal competition and/or to determine whether the concept should be proposed at all.

- *CML 6 Integrated Concept*—Expanded details on the technical, management, cost and other elements of the mission concept have been defined and documented. A NASA Step 2 CSR is at this level of maturity. There is no corresponding milestone for assigned projects.
- *CML 7 Preliminary Implementation Baseline*—Preliminary system and subsystem level requirements and analyses, demonstrated (and acceptable) margins and reserves, prototyping & technology demonstrations, risk assessments and mitigation plans have been completed. This is the maturity level needed for competed missions to hold their Preliminary Mission System Review and for assigned projects to hold their Mission Definition Review.
- *CML 8 PDR (Integrated Baseline)*—Design and planning commensurate for a Preliminary Design Review.
- *CML 9 CDR*—Design and planning commensurate for a Critical Design Review.

## ANNEX 2.B

### NASA Proposal Evaluation Factors

Tables 2.B.1 to 2.B.3 provide descriptions of NASA proposal evaluation factors for scientific merit, scientific implementation merit and feasibility, and the technical, management, and cost (TMC) feasibility of the proposed mission implementation, reproduced from NASA Standard Announcement of Opportunity Template (NASA 2018).

**TABLE 2.B.1** Evaluation Factors for Scientific Merit

| |
|---|
| Factor A-1. Compelling nature and scientific priority of the proposed investigation's science goals and objectives. This factor includes the clarity of the goals and objectives; how well the goals and objectives reflect program, Agency, and national priorities; the potential scientific impact of the investigation on program, Agency, and national science and applications objectives; and the potential for fundamental progress, as well as filling gaps in our knowledge relative to the current state of the art. |
| Factor A-2. Programmatic value of the proposed investigation. This factor includes the unique value of the investigation to make scientific progress in the context of other ongoing and planned missions; the relationship to the other elements of NASA's science and applications programs; how well the investigation may synergistically support ongoing or planned missions by NASA and other agencies; and the necessity for a space mission to realize the goals and objectives. |
| Factor A-3. Likelihood of scientific success. This factor includes how well the anticipated measurements support the goals and objectives; the adequacy of the anticipated data to complete the investigation and meet the goals and objectives; and the appropriateness of the mission requirements for guiding development and ensuring scientific success. |
| Factor A-4. Scientific value of the Threshold Science Mission. This factor includes the scientific value of the Threshold Science Mission using the standards in the first factor of this section and whether that value is sufficient to justify the proposed cost of the mission. |

SOURCE: NASA (2018).

**TABLE 2.B.2** Evaluation Factors for Scientific Implementation Merit and Feasibility

| |
|---|
| Factor B-1. Merit of the instruments and mission design for addressing the science goals and objectives. This factor includes the degree to which the proposed mission will address the goals and objectives; the appropriateness of the selected instruments and mission design for addressing the goals and objectives; the degree to which the proposed instruments and mission can provide the necessary data; and the sufficiency of the data gathered to complete the scientific investigation. |
| Factor B-2. Probability of technical success. This factor includes the maturity and technical readiness of the instruments or demonstration of a clear path to achieve necessary maturity; the adequacy of the plan to develop the instruments within the proposed cost and schedule; the robustness of those plans, including recognition of risks and mitigation plans for retiring those risks; the likelihood of success in developing any new technology that represents an untested advance in the state of the art; the ability of the development team—both institutions and individuals—to successfully implement those plans; and the likelihood of success for both the development and the operation of the instruments within the mission design. |
| Factor B-3. Merit of the data analysis, data availability, and data archiving plan. This factor includes the merit of plans for data analysis, curation, and data archiving to meet the goals and objectives of the investigation; to result in the publication of science discoveries in the professional literature; to serve and support the identified applications communities; and to preserve data and analysis of value to the science community. Considerations in this factor include assessment of planning and budget adequacy and evidence of plans for well-documented, high-level data products and software usable to the entire science community; assessment of adequate resources for physical interpretation of data; reporting scientific results in the professional literature (e.g., refereed journals); and assessment of the proposed plan for the timely release of the data to the public domain for enlarging its science impact. |
| Factor B-4. Science, exploration, or technology resiliency. This factor includes both developmental and operational resiliency. Developmental resiliency includes the approach to descoping the Baseline Science Mission to the Threshold Science Mission in the event that development problems force reductions in scope. Operational resiliency includes the ability to withstand adverse circumstances, the capability to degrade gracefully, and the potential to recover from anomalies in flight. |
| Factor B-5. Probability of investigation team success. This factor will be evaluated by assessing the experience, expertise, and organizational structure of the science team and the mission design in light of any proposed instruments. The role of each co-investigator (Co-I) will be evaluated for necessary contributions to the proposed investigation; the inclusion of Co-Is who do not have a well-defined and appropriate role may be cause for downgrading during evaluation. |

SOURCE: NASA (2018).

**TABLE 2.B.3** Evaluation Factors for the Technical, Management, and Cost (TMC) Feasibility of the Proposed Mission Implementation

| |
|---|
| Factor C-1. Adequacy and robustness of the instrument implementation plan. The maturity and technical readiness of the instrument complement will be assessed, as will the ability of the instruments to meet mission requirements. This factor includes an assessment of the instrument design, accommodation, interface, heritage, and technology readiness. This factor includes an assessment of the instrument hardware and software designs, heritage, and margins. This factor includes an assessment of the proposer's understanding of the processes, products, and activities required to accomplish development and integration of the instrument complement. This factor also includes adequacy of the plans for instrument systems engineering and for dealing with environmental concerns. This factor includes an assessment of plans for the development and use of new instrument technology, plans for advanced engineering developments, and the adequacy of backup plans to mature systems within the proposed cost and schedule when systems having a technology readiness level (TRL) less than 6 are proposed. |
| Factor C-2. Adequacy and robustness of the mission design and plan for mission operations. This factor includes an assessment of the overall mission design and mission architecture, the spacecraft design and design margins (including margins for launch mass, delta-V, and propellant), the concept for mission operations (including communication, navigation/tracking/trajectory analysis, and ground systems and facilities), and the plans for launch services. This factor includes mission resiliency—the flexibility to recover from problems during both development and operations—including the technical resource reserves and margins, system and subsystem redundancy, and reductions and other changes that can be implemented without impact to the Baseline Science Mission. |
| Factor C-3. Adequacy and robustness of the flight systems. This factor includes an assessment of the flight hardware and software designs, heritage, and margins. This factor includes an assessment of the proposer's understanding of the processes, products, and activities required to accomplish development and integration of all elements (flight systems, ground and data systems, etc.). This factor includes an assessment of the adequacy of the plans for spacecraft systems engineering, qualification, verification, mission assurance, launch operations, and entry/descent/landing. This factor includes the plans for the development and use of new technology, plans for advanced engineering developments, and the adequacy of backup plans to ensure success of the investigation when systems having a TRL less than 6 are proposed. The maturity and technical readiness of the spacecraft, subsystems, and operations systems will be assessed. The adequacy of the plan to mature systems within the proposed cost and schedule, the robustness of those plans, including recognition of risks and mitigation plans for retiring those risks, and the likelihood of success in developing any new technologies will be assessed. |
| Factor C-4. Adequacy and robustness of the management approach and schedule, including the capability of the management team. This factor includes: the adequacy of the proposed organizational structure and work breakdown structure; the management approach including project level systems engineering; the roles, qualifications, and experience of the principal investigator (PI), program manager (PM), and project systems engineer (PSE), other named Key Management Team members, and implementing organization, mission management team, and known partners; the commitment, spaceflight experience, relevant performance of the PI, PM, PSE, other named Key Management Team members, and implementing organization, mission management team, and known partners against the needs of the investigation; the prior working relationships of the implementing organization and known partners; the commitments of partners and contributors; and the team's understanding of the scope of work covering all elements of the mission, including contributions. Also evaluated under this factor is the adequacy of the proposed risk management approach, including any risk mitigation plans for new technologies, any long-lead items, and the adequacy and availability of any required manufacturing, test, or other facilities. The approach to any proposed descoping of mission capabilities will be assessed against the potential science impact to the proposed Baseline Science Mission. The plans for managing the risk of contributed critical goods and services will be assessed, including the plans for any international participation, the commitment of partners and contributors, as documented in Letters of Commitment, and the technical adequacy of contingency plans, where they exist, for coping with the failure of a proposed cooperative arrangement or contribution. This factor also includes assessment of elements such as the relationship of the work to the project schedule, the project element interdependencies, the associated schedule margins, and an assessment of the likelihood of meeting the proposed launch readiness date. Also evaluated under this factor are the proposed project and schedule management tools to be used on the project. |
| Factor C-5. Adequacy and robustness of the cost plan, including cost feasibility and cost risk. This factor includes elements such as cost, cost risk, cost realism, and cost completeness including assessment of the basis of estimate, the adequacy of the approach, the methods and rationale used to develop the estimated cost, the discussion of cost risks, the allocation of cost reserves by phase, and the team's understanding of the scope of work (covering all elements of the mission including contributions). The adequacy of the cost reserves and understanding of the cost risks will be assessed. This factor also includes an assessment of the proposed cost relative to estimates generated by the evaluation team using parametric models and analogies. Also evaluated under this factor are the proposed cost management tools to be used on the project. |

SOURCE: NASA (2018).

The AO includes additional information about how these criteria are applied to the Baseline and Threshold Science Mission, how feasibility is used in selection, and the role of student collaboration proposals. The details from the Earth Venture Mission 3 AO are included here as an example (NASA 2020a).

Scientific merit will be evaluated for the Baseline Science Mission and the Threshold Science Mission; Operational Enhancement Opportunity beyond the Baseline. Factors A-1 through A-3 are evaluated for the Baseline Science Mission assuming it is implemented as proposed and achieves technical success. Factor A-4 is similarly evaluated for the Threshold Science Mission. This evaluation will result in narrative text, including specific major and minor strengths and weaknesses, as well as an appropriate adjectival rating for the scientific merit of the investigation.

The panel evaluating the "TMC Feasibility of the Proposed Mission Implementation" may provide comments to NASA regarding the feasibility of the proposed access to space. While these comments will not be considered in the evaluation, they may be considered during selection. The panel evaluating the "TMC Feasibility of the Proposed Mission Implementation" will also provide comments to NASA regarding the extent to which the proposed investigation provides career development opportunities to train the next generation of engineering and management leaders. While these comments will not be considered in the evaluation, they may be considered during selection.

Student collaboration proposals, if any, will be evaluated only for the impact they have on the TMC Feasibility of the Proposed Investigation Implementation to the extent that they are not separable. Student collaboration proposals are not penalized for any inherent higher cost, schedule, or technical risk, as long as the student collaboration is shown to be clearly separable from the implementation of the Baseline Investigation.

# ANNEX 2.C

## Summary Evaluation Scores for Science Merit and Technical, Management, and Cost Evaluations

Tables 2.C.1 and 2.C.2 provide descriptions of NASA summary ratings for the evaluation of proposals for scientific merit and scientific implementation merit and the technical, management, and cost feasibility of the proposed mission implementation, reproduced from NASA Standard Announcement of Opportunity Template (NASA 2018).

**TABLE 2.C.1** Evaluation Rubric for Scientific Merit and Implementation

| Summary of Evaluation for Science | Basis for Summary Evaluation |
|---|---|
| Excellent | A comprehensive, thorough, and compelling proposal of exceptional merit that fully responds to the objectives of the announcement of opportunity (AO) as documented by numerous and/or significant strengths and having no major weaknesses. |
| Very Good | A fully competent proposal of very high merit that fully responds to the objectives of the AO, whose strengths fully outbalance any weaknesses. |
| Good | A competent proposal that represents a credible response to the AO, having neither significant strengths nor weaknesses and/or whose strengths and weaknesses essentially balance. |
| Fair | A proposal that provides a nominal response to the AO, but whose weaknesses outweigh any perceived strengths. |
| Poor | A seriously flawed proposal having one or more major weaknesses (e.g., an inadequate or flawed plan of research or lack of focus on the objectives of the AO). |

SOURCE: NASA (2018).

**TABLE 2.C.2** Evaluation Rubric for Technical, Management, and Cost (TMC) Feasibility

| Summary Evaluation for TMC | Basis for Summary Evaluation |
|---|---|
| Low Risk | There are no problems evident in the proposal that cannot be normally solved within the time and cost proposed. Problems are not of sufficient magnitude to doubt the proposer's capability to accomplish the investigation well within the available resources. |
| Medium Risk | Problems have been identified, but are considered within the proposal team's capabilities to correct within available resources with good management and application of effective engineering resources. Investigation design may be complex and resources tight. |
| High Risk | One or more problems are of sufficient magnitude and complexity as to be deemed unsolvable within the available resources. |

SOURCE: NASA (2018).

## ANNEX 2.D

### Categorization

The categorization process continues to focus on the scientific and technical merits, implementation feasibility, management approach, cost realism, and comprehensive risk assessment of each proposal individually. The objective of the categorization process is to derive a portfolio of consolidated merit-risk characterizations that will be subjected during the recommendation and selection processes to additional programmatic criteria (see Table 2.D.1).

**TABLE 2.D.1** Categories Used in Proposal Selection

| |
|---|
| Category I. Well-conceived, meritorious, and feasible investigations pertinent to the goals of the program and the anouncement of opportunity's objectives and offered by a competent investigator from an institution capable of supplying the necessary support to ensure that any essential flight hardware or other support can be delivered on time and that data can be properly reduced, analyzed, interpreted, and published in a reasonable time. Investigations in Category I are recommended for acceptance and normally will be displaced only by other Category I investigations. |
| Category II. Well-conceived, meritorious, and feasible investigations that are recommended for acceptance, but at a lower priority than Category I, whatever the reason. |
| Category III. Meritorious investigations that require further development. Category III investigations may be funded for further development and may be reconsidered at a later time for the same or other opportunities. |
| Category IV. Proposed investigations which are recommended for rejection for the particular opportunity under consideration, whatever the reason. |

SOURCE: NASA (2018).

# 3

# Diversity of NASA Competed Missions Teams and Space Science Workforce

Not only must the mission proposal development and review process be fair and unbiased, but a diverse proposer pool is needed to supply diverse teams for competed space missions. Several studies have demonstrated that multiple forms of diversity (i.e., cognitive, experiential, and demographic) are beneficial to the creativity, innovation, and impact of science teams (Campbell et al. 2013), provided that the challenges that often come with diverse teams are anticipated and addressed adequately (Joshi 2014; NRC 2015). Diverse teams thrive in inclusive environments where all team members feel welcomed, supported, and valued, and where mutual respect and interpersonal trust exist within the group (Hall et al. 2018). Additionally, ensuring that team members feel a sense of psychological safety (Edmondson 2004) helps create the conditions in which diverse teams perform effectively. Evidence further suggests that a critical mass of team members from historically minoritized and underrepresented groups is necessary to realize the benefits of that diversity (i.e., comprising between 15% and 30% of the team) (Cain and Leahy 2014). Though social scientists continue to try to understand how specific forms of diversity (e.g., racial/ethnic, gender, career stage diversity) relate to various aspects of team performance, it is generally understood that, when engaged productively in an inclusive environment, diverse perspectives, experiences, and backgrounds can strengthen teams, leading to better science (Sommers 2006). As noted by the Astronomy and Astrophysics Decadal Survey (Astro2020), "the pursuit of science, and scientific excellence, is inseparable from the humans who animate it" (NASEM 2021, p. 3-1). Similarly, the National Aeronautics and Space Administration's (NASA's) *Vision for Scientific Excellence* notes, "diversity is a key driver of innovation and more diverse organizations are more innovative.... NASA believes in the importance of diverse and inclusive teams to tackle strategic problems and maximize scientific return" (NASA 2020b).

## IMPORTANCE OF DEMOGRAPHICS DATA: QUALITY AND QUANTITY

Science conducted in teams is a foundational aspect of NASA's missions. Many of the science questions and engineering feats essential to addressing these questions require the broad range of knowledge, technical expertise, and disciplinary perspectives often found on science teams. To accomplish the goal of such diverse NASA mission teams, the various forms of diversity to be reflected in these teams would first need to be measured and quantified. Currently, the demographic information gathered on mission proposal teams by NASA Science Mission Directorate (SMD) is mostly limited to gender or sex, often inferred, and sometimes conflating the two categories (see Chapter 1, Box 1.3 for definitions of these terms). There is little discussion of race or ethnicity in the data provided to the committee by NASA. Additionally, while NASA's Solicitation and Proposal Integrated Review and Evaluation System (NSPIRES) does ask questions about race, ethnicity and disability status for proposers (see Annex 3.A), none of these factors were reported in a recent NASA study of demographic makeup of Astrophysics Explorer-class mission teams (Centrella et al. 2019).

The people leading proposal teams for competed missions are generally space scientists who have worked on previous missions, on scientific instruments to fly on space missions, and/or have been supported on research and analysis (R&A) grants. Thus, to understand the diversity of the leadership of competed space missions, accurate demographic data need to be collected and analyzed on the leadership and team membership of missions, on grant proposers, as well as the research workforce across the four science divisions of NASA's SMD: Astrophysics, Heliophysics, Earth Science and Planetary Science.

In this chapter, the demographics data gathered by the NASA SMD Data Analytics Team (led by Michael New, Deputy Associate Administrator for Research, NASA Headquarters) from competed mission proposals are compared with data gathered through NSPIRES which is overseen by NASA's Office of the Chief Scientist. These data are compared to the demographic data available on the space science workforce from surveys performed by professional organizations such as the American Institute of Physics (AIP) and federal agencies like the National Science Foundation, as well as with demographics of the U.S. science and engineering (S&E) workforce and overall U.S. population. SMD's two other science divisions (Biological and Physical Sciences Division and Joint Agency Satellite Division), are not included in this study because these divisions do not have competed missions. Nevertheless, the benefits of gathering comprehensive demographics data for this specific workforce likely also apply across the whole agency.

It should be noted that both the quality and quantity of data gathered by NASA and in workforce surveys are highly variable. Indeed, Astro2020 (NASEM 2021) devoted a section and two high-level recommendations directing the agencies (including NASA) to systematically and deliberately collect and analyze demographics data across the portfolio of activities (see Astro2020 Section 3.3.6). Annex 3.A provides examples of the questions asked by NASA and AIP to collect demographic information as well as a discussion of the issues involved. See Chapter 6 for comparison with the methods applied by other agencies to gather such data and lessons to be learned therefrom.

## DEMOGRAPHICS OF MISSION PROPOSERS ACROSS THE FOUR SMD SCIENCE DIVISIONS

As discussed in Chapter 1, the first AO for a fully competed and totally PI-led mission (as we know them today) was the Planetary Division's Discovery AO released in 1994. For the first few years, the number of opportunities were limited. By 2006 most divisions were regularly issuing AOs for competed missions. Since 1996, there have been 701 mission proposals submitted to NASA, of which 143 were selected.

Over the same period, there were 232 proposals submitted for instruments on flagship missions (usually led by a NASA center) of which 46 were selected. The NASA SMD Data Analytics Team have examined these 933 proposals to gather statistics on how demographics of the proposal teams varies with SMD division, size of mission, and year of submission (Jensen and Pappas 2021). For this report, the scope is limited to proposals submitted in the past decade (2010-2019) since the primary interest is the current proposal process. Table 3.1 shows the number of proposals submitted during this time period for missions of various sizes (categorized by maximum cost) plus instruments for flagship missions. Missions of opportunity (MOs) can be an open call for hardware—a small, development spacecraft or an instrument, or contribution to a non-U.S. mission.

It is important to note that the (binary) gender—woman/man (W/M)—of mission proposal personnel were derived via two steps: (1) the NSPIRES system provided full names, team member roles, and institutions for all proposal personnel; (2) genders of team members were obtained from their first names using the online tool Gender API[1] (team members having the first (given) names with reported accuracy of 95% or higher for inferred gender were assigned to the binary W/M categories; otherwise, binary gender was determined using public domain

[1] See https://gender-api.com, which uses first names to infer binary woman/man gender from a database of over 2.1 million names, covering 178 countries.

**TABLE 3.1** Submissions and Selections Between 2010 and 2019 Across SMD and for PIs Inferred to Be Women (W PI) for Different Mission Sizes

| Mission Size | Cost Cap | Number of AOs[a] | Number of Proposals | %W PI Proposed | Number Selected | % Selected | %W PI[b] Selected |
|---|---|---|---|---|---|---|---|
| MO/Instrument | <$125 million | 21 | 326 | 11% | 65 | 20% | 12% |
| Small | <$250 million | 5 | 89 | 4% | 18 | 20% | 0% |
| Medium | <$750 million | 5 | 97 | 23% | 24 | 25% | 46% |
| Large | >$750 million | 1 | 12 | 42% | 4 | 33% | 50% |
| Mission Total | | 11 | 198 | 16% | 46 | 23% | 28% |
| Overall Total | | 32 | 524 | 13% | 111 | 21% | 19% |

NOTE: MO = Mission of Opportunity.

[a] Some announcements of opportunity include more than one mission size.

[b] Percentage of selected proposals with woman as PI.

SOURCE: Heidi Jensen and Lorenzo Pappas, NASA Science Mission Directorate.

searches). As it relates to career stage for team members, public domain searches of curricula vitae (CVs) were made to determine the number of years since PhD—a proxy for career age—for each member. For all discussions in this chapter, only de-identified versions of these data were provided to the committee.

### Award Rates

The average rates at which mission proposals (across all of SMD) are selected for funding range from 20% to 33%, depending on mission size. The percentage of proposals submitted by a woman PI range from 4% (for small missions) to 42% (for large missions) while the percentage of successful proposals with a woman PI range from 0% (for small missions) to 50% (for large missions) as shown in Table 3.1. To further explore this wide range of success rates, particularly for women PIs, the proposals were categorized according to the specific SMD division to which they were submitted. Table 3.2 shows the distribution of 524 proposals submitted (2010-2019) across the four SMD science divisions for different mission sizes, separated by inferred gender of the PI and whether funded or not. The total number of proposals in each mission class were similar for Astrophysics (total 101, 7% women PIs) and Heliophysics (total 96, 8% women PIs). Earth Science had only small missions and MOs (total 108, 5% women PIs) in this time period, and in Planetary Science there were no small missions (which are usually limited to Earth orbit) in this time period, while there were a number of instrument and medium missions (total 219, 22% women PIs).

Most notable is that there were no women PIs funded in Astrophysics and Earth Science during this time-period (see Table 3.2), while in both Heliophysics and Planetary Science ~38% of proposals submitted by women were

**TABLE 3.2** Number of Proposals Funded and Denied Between 2010 and 2019 for Each NASA Science Mission Directorate Science Division and Inferred Gender of Principal Investigator

| Total 524 | Astrophysics 101 | | Earth Science 108 | | Heliophysics 96 | | Planetary Science 219 | |
|---|---|---|---|---|---|---|---|---|
| | Funded | Denied | Funded | Denied | Funded | Denied | Funded | Denied |
| Men | 22 | 72 | 11 | 92 | 30 | 58 | 27 | 144 |
| Women | 0 | 7 | 0 | 5 | 3 | 5 | 18 | 30 |
| Total | 22 | 79 | 11 | 97 | 33 | 63 | 45 | 174 |

SOURCE: Heidi Jensen and Lorenzo Pappas, NASA Science Mission Directorate.

funded. Using Fisher's exact test (Mehta and Patel 1983) for count data, preferred in situations when cell counts are small, the null hypotheses that funding rates are the same for men and for women are tested for each SMD division separately. For Astrophysics, Earth Science, and Heliophysics, all respective p-values were greater than 0.3. That is, considering all submissions to these divisions over 2010-2019, Fisher's exact tests show no evidence that funding rates vary by gender in Astrophysics, Earth Science, and Heliophysics. Note, however, that even if there were differences in funding rates, detecting them with statistical significance is impossible due to the small number of proposal submissions from women PIs in these three divisions. For example, in Astrophysics, given a total of 94 submissions by men PIs with 22 of those being funded, detecting a statistically significant (at the 0.05 level) difference in funding rates by gender would only be possible if there were 15 or more total submissions by women PIs. For Planetary Science, on the other hand, the overall number of proposal submissions between 2010-2019 is about twice as large, and the number of submissions from women PIs is at least five times as large as in other divisions, and the p-value of Fisher's exact test is 0.002. Thus, in Planetary Science, between 2010-2019, the funding rate for women was significantly greater than the funding rate for men. In fact, if the funding rates were the same, only about 10 women out of 48 that submitted competed mission proposals as PIs in Planetary Science would have been expected to be funded. In the analysis of the data, it is noted that there were 18 women PIs funded in Planetary Science between 2010-2019 (see Table 3.2).

The above analysis is presented for illustrative purposes and is accompanied by several caveats in interpretation. In addition to difficulties detecting statistical significance due to small numbers of submissions by women PIs in three out of four SMD divisions, the analysis is also limited in other aspects. First, the aggregated summary does not reflect important dependencies on, for example, mission size, submission year, or specific AO. These are all important confounders and there are likely to be others. Some parameters, such as mission size, were available to the committee, while other specifics such as individual AO and submission year were not. Second, the same individual PIs could have had multiple submissions among funded, denied, or both denied and funded proposals counted in Table 3.2, and PI individual identifiers were not available. Additionally, significant associations can disappear, get stronger, or reverse the direction when data is disaggregated (Blyth 1972). While appropriate choice of the control variables is an important part of any research study (Agresti 2003) it is beyond the scope of this committee's task.

Additionally, the lack of statistically significant differences by gender should not be interpreted as an indication of a lack of gender bias in the review process. Assessments of the presence or lack of gender bias in these decisions would require more information about the qualifications and experience of the PIs in question, and no such data are available. Research on gendered patterns in science, technology, engineering, and mathematics (STEM) specifically, and gendered labor force patterns more broadly, demonstrates that women are often less likely than men to consider themselves qualified for funding, awards, or job applications, even when they have identical credentials (Waisbren et al. 2008; Moss-Racusin and Rudman 2010; Goulden et al. 2011; Mijid 2014). If the women PIs were more qualified than the men PIs who submit mission proposals, then the lack of significant gender patterns would indicate gender bias because women would have been held to a higher standard than men to have the same chances for an award (Pohlaus et al. 2011). In short, identifying the existence or absence of gender bias in the proposal assessment process would require in-depth analyses involving proposal-specific and PI-specific information.

The committee acknowledges that there is no easy solution that would ensure unbiased and fair allocation of competed mission funding with respect to gender, racial, ethnic, and other historically marginalized groups within the complicated and high stakes proposal development and review process employed by NASA (see Chapter 2). Because of the ways in which social group identities can be tied to research qualifications and experience, it is crucial to continuously evaluate the process and compare with current research about social disparities in science's interwoven reward systems, as well as involve experts who specialize in the complicated ways in which disparities associated with historically marginalized groups manifest themselves in society (Lee et al. 2020).

### Pathways

To understand the career pathway to becoming a PI of a competed mission, the academic age of PIs (categorized as the number of years since PhD of the PI at the time of proposal submission) is assessed for all proposals submitted to NASA SMD between 2010 and 2019. Most submitted proposals have a PI who completed their PhD >25 years before proposal submission (see Figure 3.1; top plot). Most notable in the distribution of academic age for PIs of accepted proposals is that the profile is flatter for Astrophysics and Earth Science, while for Heliophysics, and, to a lesser extent, for Planetary Science, the PIs of successful proposals are predominantly older men. Overall, the data suggests that women PIs tend to be younger.

If one takes age 30 as the typical age for completing a PhD, these data suggests that mission PIs are typically >55 years of age. Currently 20% of U.S. PhDs in physics and 40% of U.S. PhDs in astronomy are awarded to women (Porter and Ivie 2019). We focus on physics and astronomy given an appreciable proportion of mission proposers earn doctorates in these fields. Looking back 30 years (~1990) the fraction of physics and of astronomy PhDs awarded to women was 8% and 18% respectively (see Figure 3.2). These percentages are consistent with the percentage of women PIs of proposals in Heliophysics, where PIs are likely to have earned a PhD in physics >25 years ago. While such trends provide some explanation for the low representation of women PIs, they do not explain the stark variation across SMD divisions.

Further analysis of the data on proposals selected for funding demonstrated that both the number of accepted mission proposals and the rate of acceptance varies between divisions, depending on mission size and on inferred gender (see Figure 3.3). The committee acknowledges that different divisions will need missions of different sizes to address their very different science goals, yet, the lack of women PIs, both proposing and being funded, in Astrophysics and Earth Science for almost an entire decade is a significant gap. Overall acceptance rates in the 20-30% range are consistent with the average acceptance rate across all SMD divisions (see Table 3.1), except for Earth Science where 10% is more typical. Additionally, the acceptance rates for women in Heliophysics are relatively comparable to men (except for non-involvement of women PIs in small missions) and the acceptance rates for women are higher in Planetary Science. These wide-ranging demographics of mission leadership beg the question about what differences in the pathways, workforces, proposal processes and/or cultures of these divisions lead to such variation. Addressing these questions at the specific divisional level could not be accommodated in this report without compromising confidentiality, but could certainly be taken up within the context of the NASA-funded decadal surveys that are carried out for each of the SMD divisions and also involve a larger pool of Earth and space scientists for the purposes of examining the state of the profession. See Chapter 5 for a discussion of factors that may contribute to disparate outcomes within STEM fields more broadly.

### Mission Teams

For many mission PIs, an important step in their path to becoming a PI is involvement on a mission team, often as a co-investigator (Co-I). Chapter 2 (see Figure 2.8) shows the distribution of roles held by members of the science teams of competed missions (2006-2019). The majority (81%) of team members are Co-Is. A Co-I (or Collaborator) usually provides a limited, focused component to the project. A few mission teams are recognizing the importance of training future Co-Is and PIs as well as including students and/or postdoctoral researchers on the team (see Chapter 4). Correspondingly, the distribution of academic age amongst team members is shifted about 10 years younger than the academic age distribution of PIs shown above.

Over the past decade, the data from the NASA SMD Data Analytics Team for 2010-2019 show that the total number of Co-Is were 3,414 in Planetary Science versus 1,640 in Heliophysics, 1,452 in Earth Science, and 1,343 in Astrophysics. Dividing these numbers by nine results in the total number of Co-Is submitting per year to be 379 for Planetary Science, 182 for Heliophysics, 161 for Earth Science, and 149 for Astrophysics.

Figure 3.4 shows the distribution of proposals (2010-2019) that were submitted (left) and accepted (right) for the four science divisions versus the size of the team and sorted by (inferred) gender of the PI. No analogous information was available to the committee regarding PIs race or ethnicity. Comparing the distribution of team

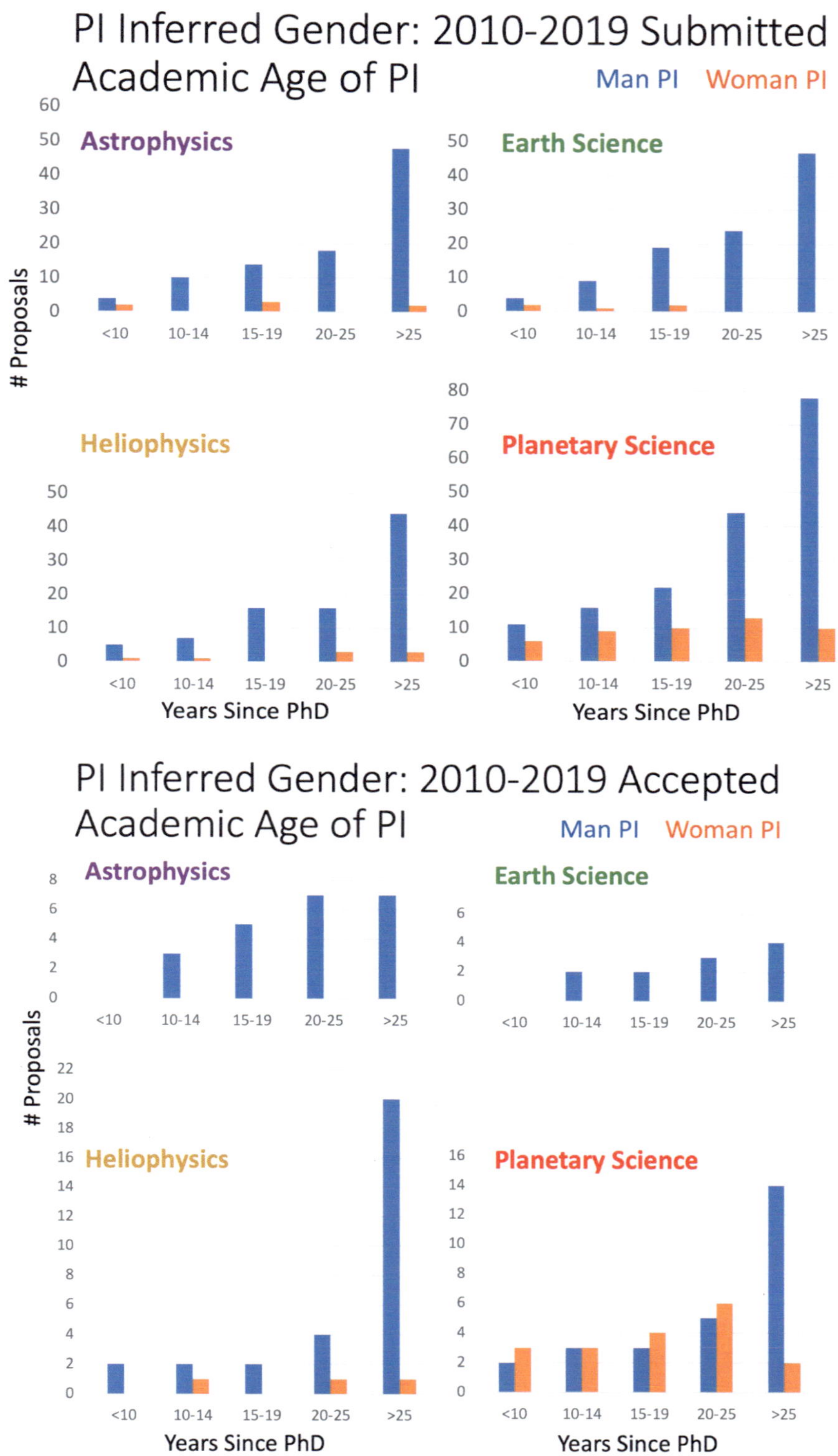


**FIGURE 3.1** Number of mission proposals (*top*) submitted and (*bottom*) accepted between 2010 and 2019 by PIs for each of the four science divisions, sorted by academic age (defined as years since PhD) and by inferred gender. Note the scales are similar between divisions but different left and right. SOURCE: Based on data from Heidi Jensen and Lorenzo Pappas, NASA Science Mission Directorate.

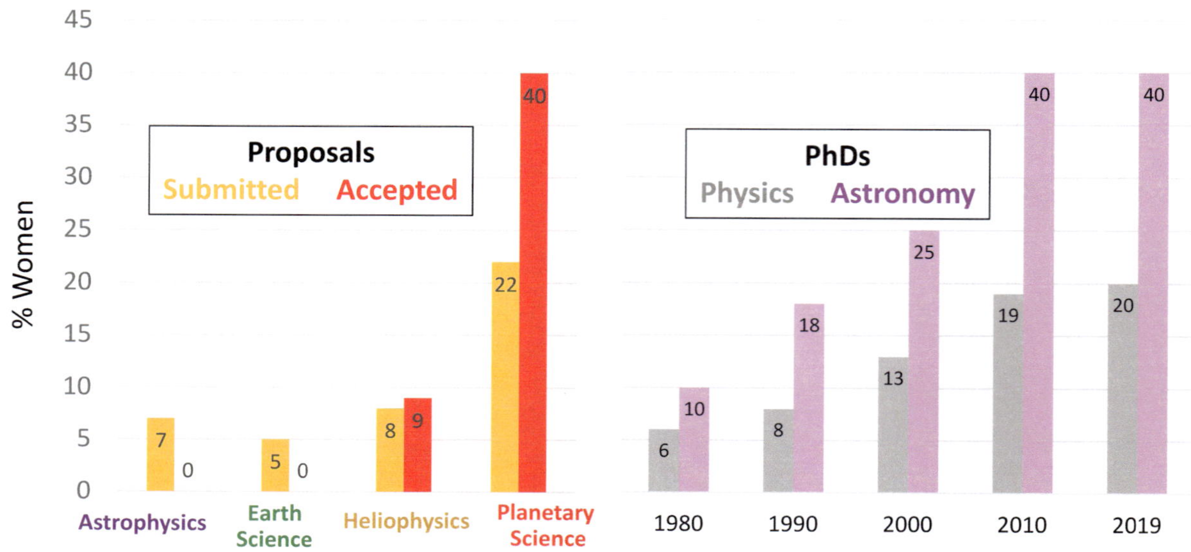


**FIGURE 3.2** (*Left*) Percentage of all mission proposals (including MO and instruments) submitted and accepted between 2010 and 2019 by women PIs for each of the four science divisions, compared with (*Right*) percentage of PhDs in physics and in astronomy awarded to women over the past 40 years. SOURCES: (*Left*) Heidi Jensen and Lorenzo Pappas, NASA Science Mission Directorate; (*Right*) Porter and Ivie (2019). See https://www.aip.org/statistics/reports/women-physics-and-astronomy-2019.

sizes for accepted versus submitted proposals suggests that perhaps larger teams are proportionally more successful, particularly in Astrophysics and Heliophysics, though the small-number statistics may not provide solid evidence.

The involvement of women in proposed mission science teams has been slowly increasing with time across SMD. The top graph of Figure 3.5 shows the average percentage of women (inferred) on science teams on submitted proposals for each year from 2006 to 2019 (the only range of data provided to committee). The gray line connects the points of the yearly percentage women participation in science roles across all submitted proposals over all of SMD. Averaged across SMD and across team size, the participation of women has increased over the 13 years from ~10% to ~25%. The colored symbols (top, Figure 3.5) correspond to the respective average yearly women participation observations in each division. Since Planetary Science greatly outnumbers the other SMD divisions (e.g., see Figure 3.7 in the next section), the average line tends to closely follow the yearly observations for women participation in Planetary Science proposed mission science teams. Note, however, that the participation of women in Planetary Science mission teams has been above the average after 2010, rising to ~33% in 2019. The participation of women in Astrophysics and Heliophysics roughly follow the trend across SMD. By contrast, in Earth Science the mission participation of women has remain quite flat between 12-15%.

The bottom graph of Figure 3.5 shows the composition of mission teams—specifically, the number of (inferred) women on a team versus size of the team for proposals submitted between 2006 and 2019. The red lines illustrate corresponding percentage of (inferred) women on the teams. We note that there are a number of missions in all SMD divisions where there are no women on the team (illustrated by red dots along the x axis). For all but a few proposals, the participation of women on mission teams is less than 30% across all divisions except Planetary Science where there are several with 30-50% participation by women and even one proposal with 73% women. It would be useful to compare these statistics for proposal submissions with the corresponding data from the proposals that were funded. Moreover, given these data plots were provided to the committee with the inclusion of statistics from 2006-2010, further study is needed to determine the specific influence of these earlier years on the current distributions.

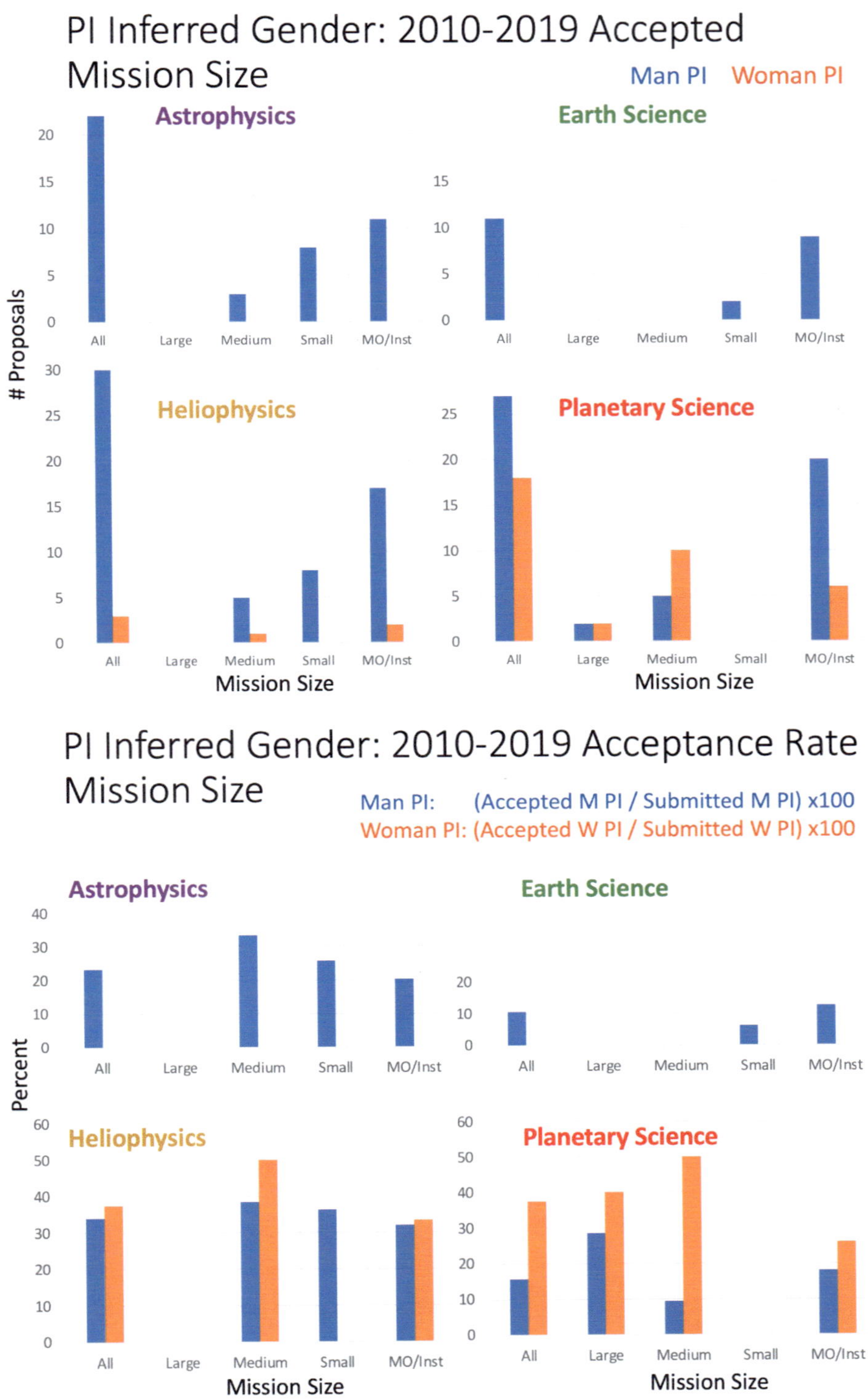


**FIGURE 3.3** (*Top*) Number of mission proposals accepted between 2010 and 2019 for PIs in each of the four science divisions, sorted by mission size and by inferred gender. (*Bottom*) The percentage acceptance rate of mission proposals, sorted by mission size and by inferred gender, noting that for Astrophysics and Earth Science the numbers for women are 0/0. The scales are similar between divisions but different left and right. SOURCE: Based on data from Heidi Jensen and Lorenzo Pappas, NASA Science Mission Directorate.

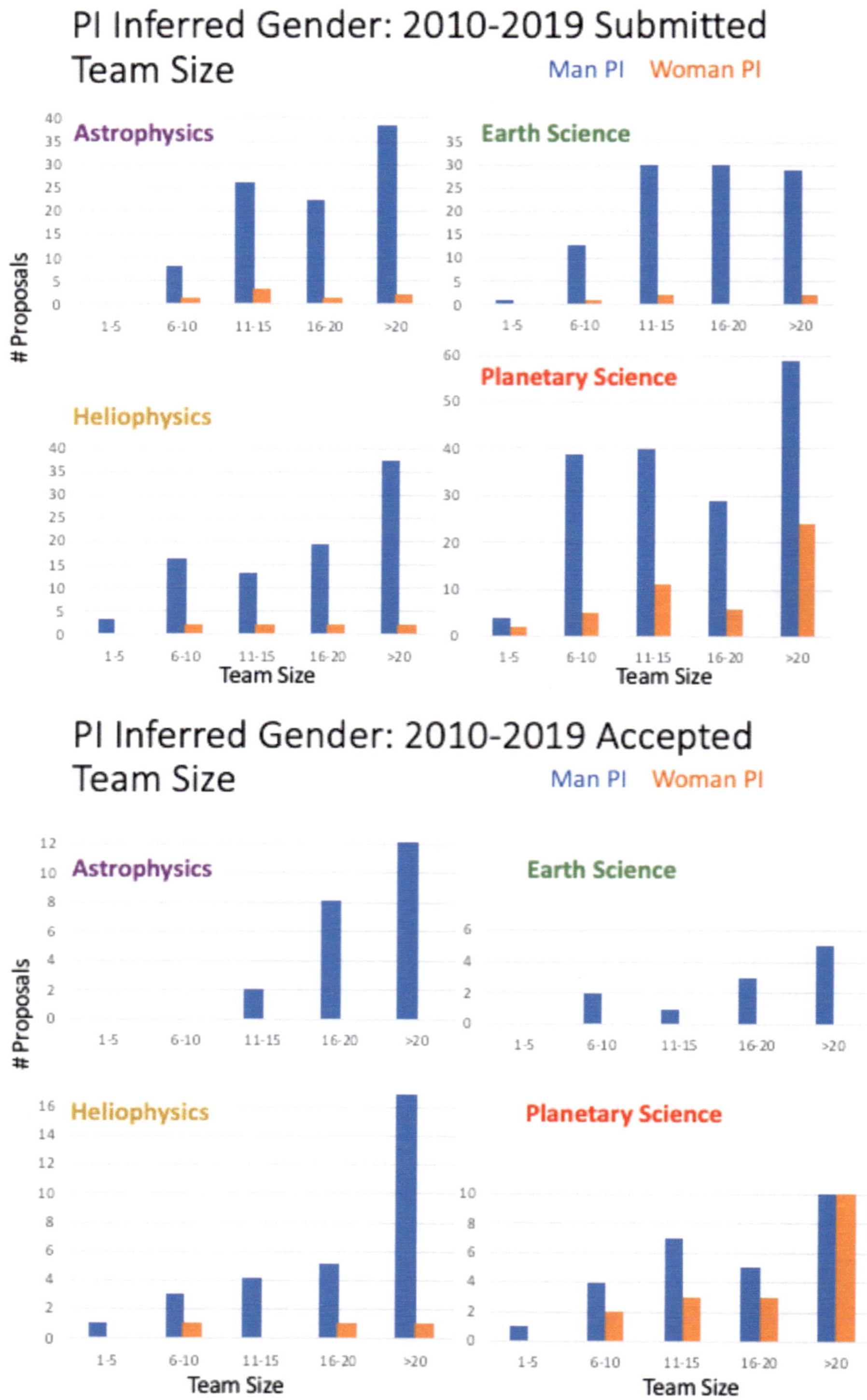


**FIGURE 3.4** Number of mission proposals (*top*) submitted and (*bottom*) accepted between 2010 and 2019 for PIs in each of the four science divisions, sorted by team size and by inferred gender. Note the scales are similar between divisions but different left versus right for submitted vs accepted proposals. SOURCE: Based on data from Heidi Jensen and Lorenzo Pappas, NASA Science Mission Directorate.

**Conclusion 3-1:** The data gathered by NASA on the demographics of proposal teams, both competed missions and R&A grants, is inadequate to inform NASA about the diversity of space scientists funded by the agency. In particular, while some data were presented on (binary) gender identity of members of proposal teams, information about race, ethnicity, and disability are lacking in NASA's analyses.

**Conclusion 3-2:** During 2010-2019, fewer than 10% of all competed mission proposals to the Astrophysics, Earth Science, and Heliophysics divisions of NASA SMD were submitted by women PIs. Zero women-led competed missions were selected for funding in both the Astrophysics and Earth Science divisions. The low number of submissions make it impossible to determine whether there are statistically significant gender differences in funding rates for these divisions. Over the same period, 22% of all competed mission proposals to the Planetary Science division were submitted by women PIs, and women-led competed missions were selected at a significantly higher rate than men-led competed missions in this division. No analogous data on PI race or ethnicity were available to the committee.

**Conclusion 3-3:** Without further in-depth analyses involving proposal-specific and PI-specific information, the absence of evidence of gender differences in aggregate funding rates over 2010-2019 cannot be interpreted as an indication of the presence or absence of gender bias in NASA proposal review process.

## DEMOGRAPHICS OF ALL PROPOSERS ACROSS THE FOUR SMD SCIENCE DIVISIONS

To put the demographics of competed mission proposals in a larger context, this section considers demographics data gathered on all proposals submitted to the science divisions of SMD. The NASA Solicitation and Proposal Integrated Review and Evaluation System (NSPIRES) was implemented in 2006, and is the primary online proposal submission interface at NASA for programs that encompass all of the competitively funded research for NASA Earth and Space Sciences, ranging from research projects on the order of $100,000 per year to the >$750 million large mission proposals. The programs support researchers at all career levels, from new investigators at the graduate student level (Future Investigators in NASA Earth and Space Science and Technology [FINESST]), the Early Career Investigators programs, extending to large, interdisciplinary research (Interdisciplinary Research Science [IDS]) to competed instrument and mission proposals. Beginning in 2016, NSPIRES users, including proposers, reviewers, and students, were asked to create personal profiles based on a series of nine questions (see Annex 3.A).

The data collected via NSPIRES consisted of gender, ethnicity, race, and disability/serious health conditions. It also asked if the person had ever been a PI, program director (PD), Co-PI, or Co-PD on any federally funded project (not specific to NASA). In 2019, NASA added an additional option for the "gender" question, specifically "Other" (rather than only "Male" or "Female"). Note that the question intended to collect data about gender actually asks about biological sex. While gender and biological sex are conflated in the original data provided to the committee, in this report, all respective data charts and tables are labeled as gender. In 2019, questions were also added on "Career Status," including Highest Degree Achieved, Year of Highest Degree, Career Classification Sector, and Career Type. NSPIRES requests but does not require responses to these questions, and "I prefer not to answer" (PNA) is one of the possible options. It is important to note that data are not recollected each time an NSPIRES user visits the site, nor are users requested to update their profile. The current list of questions asked under the NSPIRES Personal Profile is provided in Annex 3.A, but is in the process of being updated to include more expansive categories for identification (e.g., binary and non-binary gender, disability type, etc.) among other changes.

To gain some insight into the potential community of mission proposers, the demographic data collected via NSPIRES on all research and analysis (R&A) grant proposers and disaggregated by SMD science division by NASA's Office of the Chief Scientist were used as a proxy in this report (Barbier et al. 2021). As another caveat, we note that demographic data were collected beginning in 2016 and, where possible, the data were "back-casted" for proposals submitted in 2014 and 2015. Figure 3.6 shows the gender distribution (for both PIs and Co-Is combined)

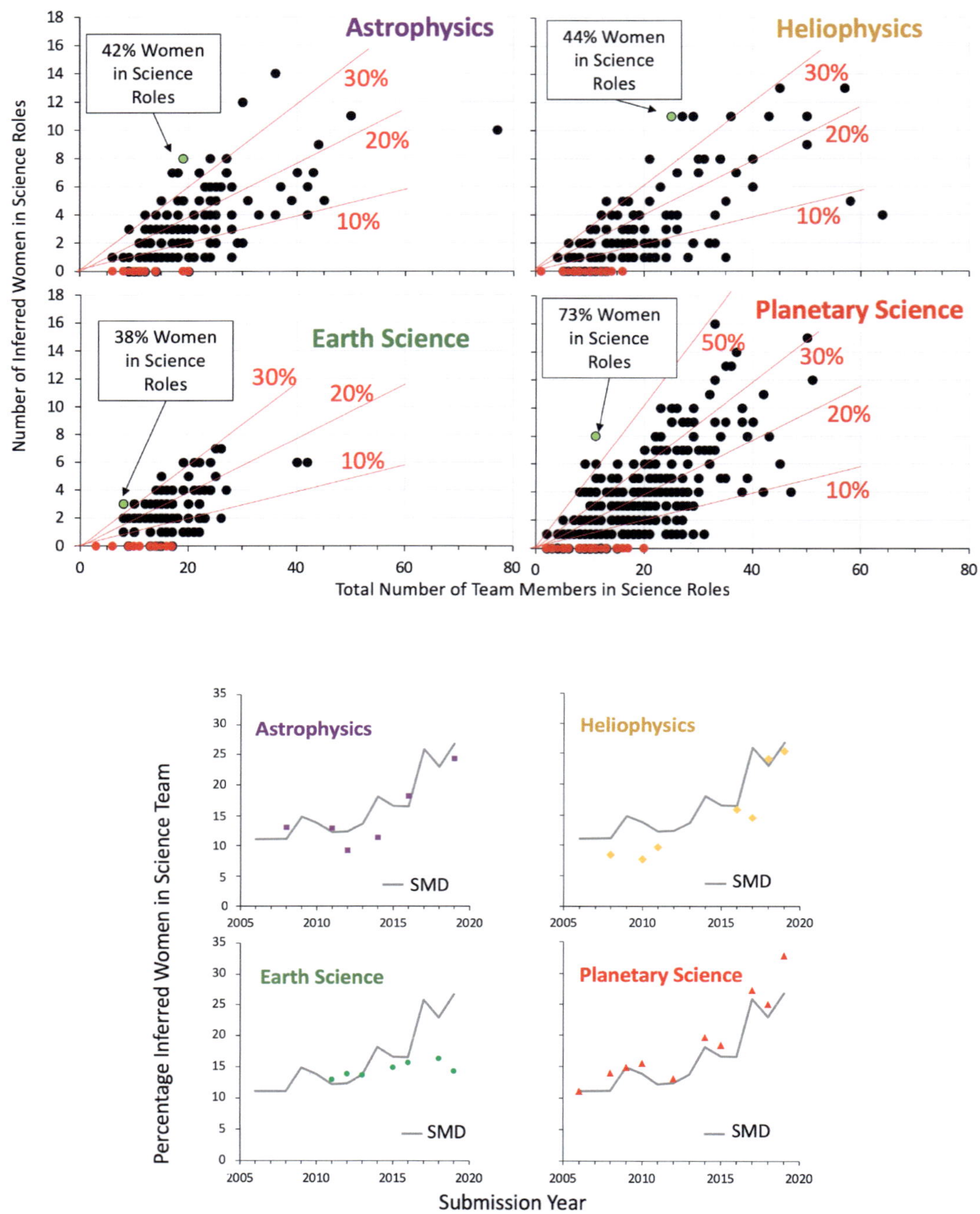


**FIGURE 3.5** Participation by women in science roles on competed missions and MO/instrument proposals submitted between 2006 and 2019 across the four science divisions that fund PI-led missions. (*Top*) Percent participation of (inferred) women in science roles on competed mission proposals (averaged over the year) plotted versus proposal submission year. In each of the four plots the gray line shows the percentage women participation in science roles across all submitted proposals over all of SMD. (*Bottom*) Number of (inferred) women in science roles compared with the total team size. The red dots show teams with zero women in science roles, while the green dots reflect teams with the highest participation of women in science roles for each division. Note that there are sometime multiple points on top of each other. SOURCE: Based on data from Heidi Jensen and Lorenzo Pappas, NASA Science Mission Directorate.

across proposals submitted between 2014 and 2020 in each science division. Across the four divisions, there were similar percentages proposals submitted by men (65-72%), women (19-23%) and by those who selected "prefer not to answer" (9-14%). For Planetary Science, the percentage of people (0.1%) who chose "other" for gender is also included. Note that the proportions of submissions with unknown ("prefer not to answer") PI and Co-I gender are substantial.

In Figure 3.6, for each division, the total number (N) of Investigators includes multiple counts of the same individuals (PIs and Co-Is) if they participated in multiple proposals. Table 3.3 shows these data disaggregated by SMD science division. When the total number of Investigators (PIs and Co-Is) is divided by the total number of proposals submitted, the average number of Investigators per proposal ranges from 3.2 to 4.2 across the divisions. When the total number of proposals submitted are averaged for each year, there are notable differences between the divisions, with Earth Science and Planetary Science submitting roughly twice the number submitted by the Astrophysics and Heliophysics divisions. When the unique names of PIs are considered, it is possible to evaluate how many proposals the PIs submitted per year. The resulting number of proposals submitted per PI does not vary much for each division across the 7 years.

Considering the most recent data (2020), the percentage of PIs who submitted just one proposal in 2020 ranges from 74% for Earth Science to 90% for Astrophysics. The percentage of PIs who submitted two proposals in 2020 ranged from 8.5% in Astrophysics to 19% in Earth Science. The remaining percentage of PIs who submitted more than 2 proposals ranged from 1.5% for Astrophysics to 7% for Earth Science. These data give some indications of the amount of effort expended by scientists in submitting research proposals to NASA but to fully understand the implications across the workforce, further analysis is required (such as submission versus selection rates and across dimensions of gender and race/ethnicity).

Within NSPIRES, there are limited data available on the race and ethnicity of proposers. Figure 3.7 shows the participation rates of various racial/ethnic groups (PIs and Co-Is combined) as Investigators across the four divisions. Due to small numbers of respective Investigators, the following racial/ethnic groups are combined into "Other": American Indian or Alaska Native, Multi-racial, Native Hawaiian or Other Pacific Islander, and Other. The racial/ethnic distribution is very similar across the four divisions, and participation is heavily dominated by investigators who identify as White (56-66%). The participation rates of Investigators who identify as Asian is ~9-20%, while the percentages of Investigators who identify as Black/African American, Latinx/Hispanic, or Other are very small (~0.4-5%). Again, the large percentage of Prefer-Not-to-Answer responses (~18-22%) could have a significant impact on the statistics if there is a tendency of a particular racial/ethnic group that prefers not to identify.

As discussed in the next section, the numbers of active researchers (those receiving research funding from NASA and/or the National Science Foundation [NSF]) in each of the four SMD divisions are comparable, and divisional differences in proposals submitted to NASA may be related to a significant fraction of funding coming from NASA versus NSF or other agencies. Results from workforce surveys in the space sciences show that for Astronomy, the percentage of respondents receiving funding from NASA was 68% and from NSF was 36%, with many scientists getting funding from both agencies. The average percent of total funding reported by astronomers from NASA was 72% and from NSF was 57%. For Heliophysics, of those getting support external to their employer institution, 74% said the funding came from NASA, 46% said from NSF. The median level of support for Heliophysics research from NASA was 75% and from NSF was 50%. In Planetary, non-university employees were asked about their source of funding, to which 54% respondents reported funding from NASA grants, 43% from NASA mission support, and only 5% reported funding from NSF. The relative funding rates from NASA versus NSF are not available for the Earth observing community of investigators.

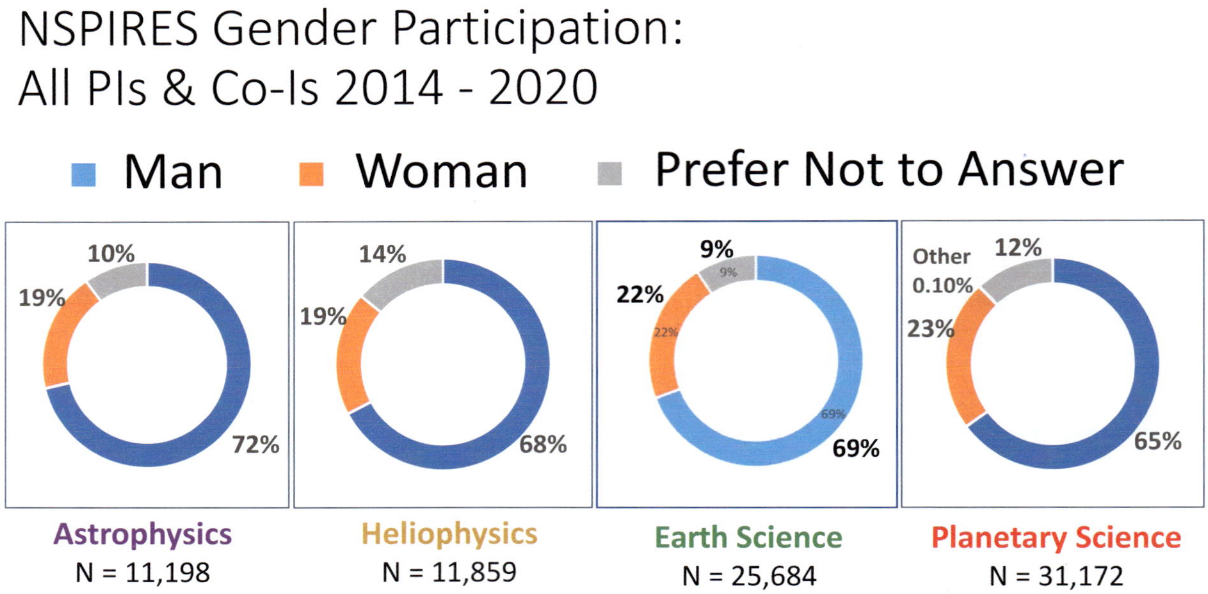


**FIGURE 3.6** Percentage participation on research and analysis proposal teams submitted via NSPIRES for divisions of SMD, sorted by gender as entered into Personal Profile by each team member. Co-Is can include graduate students. The number N is the total number of PIs plus Co-Is for each division over the 2014-2020 timeframe. SOURCE: Office of the Chief Scientist, NASA Headquarters.

**TABLE 3.3** Proposals Submitted to NSPIRES Across NASA Science Mission Directorate Science Divisions

| Total | Astrophysics | | Earth Science | | Heliophysics | | Planetary Science | |
|---|---|---|---|---|---|---|---|---|
| 524 | 101 | | 108 | | 96 | | 219 | |
| | Funded | Denied | Funded | Denied | Funded | Denied | Funded | Denied |
| Men | 22 | 72 | 11 | 92 | 30 | 58 | 27 | 144 |
| Women | 0 | 7 | 0 | 5 | 3 | 5 | 18 | 30 |
| Total | 22 | 79 | 11 | 97 | 33 | 63 | 45 | 174 |

NOTE: Data for 2020 were not complete.
SOURCE: Office of the Chief Scientist, NASA Headquarters.

## DEMOGRAPHICS OF SPACE SCIENCE WORKFORCE

To understand the demographic profile of the space sciences workforce in the United States, the committee relied on workforce surveys carried out by professional organizations, such as the American Institute of Physics (AIP). The AIP is an umbrella organization that encompasses organizations such as the American Physical Society, the American Geophysical Union (AGU), and the American Astronomical Society (AAS); and includes a Statistical Research Center with experience in surveying the workforce of the corresponding academic fields. As explained below, such surveys have their limitations, but provide estimates of the total numbers of research scientists working in the United States and their demographics. To date, there has been no systematic survey of the workforce conducted by NASA SMD to evaluate the total size and overlap between divisions, let alone their demographics or stark differences between divisions.

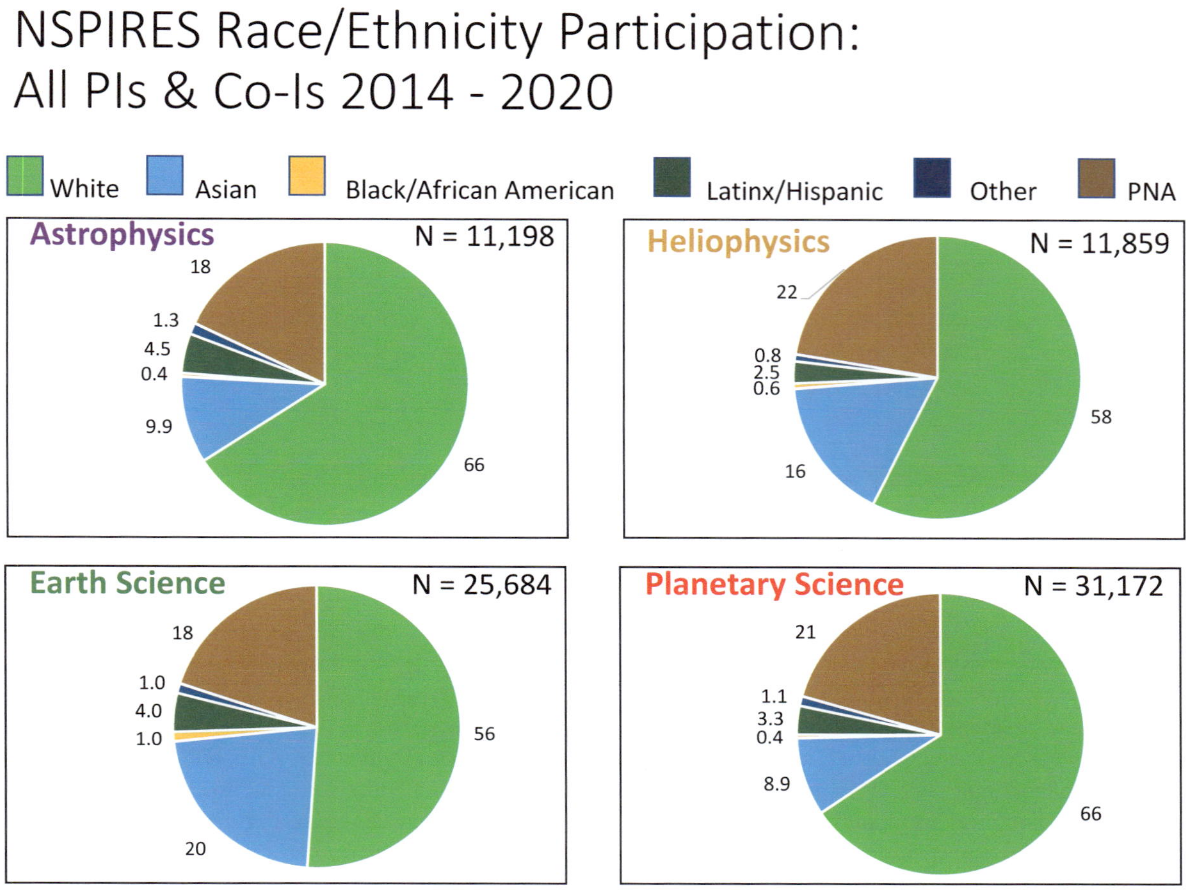


**FIGURE 3.7** Participation as PI or Co-I on all proposals submitted via NSPIRES between 2014 and 2020. Race/ethnicity information is submitted via questions on the Personal Profile questionnaire. The values of N for each division are the total numbers of proposals submitted. SOURCE: Adapted from Office of the Chief Scientist, NASA Headquarters.

### Astrophysics

The most recent workforce survey in astrophysics was conducted in 2018 by the AIP in preparation for the National Academies' Decadal Survey on Astronomy and Astrophysics 2020, and the survey results reflect demographic data for members of the AAS (Pold and Ivie 2018). The AAS is the primary affiliation of professional astronomers in the United States, and while a significant proportion of astronomers are not involved in NASA space missions (e.g., ground-based observations and theoretical research funded by NSF), the demographics reported across AAS are a reasonable representation of the workforce associated with NASA's astrophysics division. The AIP survey was sent to the 3,354 AAS members in the United States, with a response rate of about 60% (n=2018 responses). The (biological sex and race/ethnicity demographics of this population is shown in Figure 3.8. Based on the number of respondents who were PhD recipients (n=1,431), a rough "ballpark" estimate can be made that the total U.S. post-PhD workforce in astrophysics is approximately 2,400.[2]

[2] The caveat is noted for these "ballpark" numbers that the population who responded is assumed to be the same as the population who did not respond.

### Heliophysics

The most recent workforce survey in heliophysics was conducted in 2011 by the AIP (White et al. 2011a) in preparation for the *Solar and Space Physics: A Science for a Technological Society* decadal survey (NRC 2013), and the survey results reflect demographic data for Solar, Space & Upper Atmospheric Physicists identified across various sources: the AGU Space Physics and Aeronomy Section, the AAS Solar Physics Division, Space Weather Week attendee lists, and NSF PI lists. This AIP survey was sent to 2,560 unique email addresses, with a response rate of about 51% (n=1,305 responses). Of the responses received, 1,171 respondents indicated that they considered themselves in the field of Solar, Space and Upper Atmospheric Physics and currently work and reside in the United States. The biological sex and race/ethnicity demographics of this population is shown in Figure 3.8. Based on the number of respondents who were PhD recipients, a rough "ballpark" estimate can be made that the total U.S. post-PhD workforce in Heliophysics is approximately 2,300.

### Planetary Science

The most recent workforce survey in planetary science was conducted in 2020 by the AIP in preparation for the National Academies' Planetary Science and Astrobiology Decadal Survey 2023-2032, and the survey results reflect demographic data for members of the AAS Division of Planetary Science (DPS) (Porter et al. 2020). Planetary science is a particularly interdisciplinary field of research with scientists belonging to a range of professional organizations and attending diverse conferences. Furthermore, the field is expanding into astrobiology and exoplanets. Thus, the planetary science workforce likely has the most overlap with other science divisions. The 2020 AIP survey was sent to 4,965 members of the planetary science community across the AAS-DPS, the planetary division of the Geological Society of America, and attendees of the Lunar and Planetary Science Conference (LPSC), with a response rate of about 48% (n=2,367 responses). It is important to note that the AGU declined participation in the 2020 workforce survey, though the preceding workforce survey of members of AAS-DPS, AGU-planetary and LPSC, showed that only 12% of respondents belonged only to the AGU (White et al. 2011b). The biological sex and race/ethnicity demographics of the planetary science population is shown in Figure 3.8. Again, a rough "ballpark" estimate can be made that the total U.S. workforce in planetary sciences is approximately 1,960.

### Earth Science

NASA's Earth Science Division specifically involves space-based observations of Earth, and such observations look at land, ocean, and the atmosphere. To date, no workforce survey has been conducted for this sub-field of Earth science alone, and there is no single professional organization of which such Earth-observers are likely to be members. To get a sense of the U.S. Earth Science workforce as it relates to NASA space missions, the committee relied on two sources: the 2018 NSF Report of Earned Doctorates in the United States (NSF 2018) and the work of Bernard and Cooperdock (2018). The latter looks at the demographic data on geoscience PhDs included in the aforementioned NSF report and further analyzes these data by sub disciplines—Earth, ocean, and atmospheric sciences. In this consensus report, the focus is placed on the demographic data in the atmospheric sciences presented in Bernard and Cooperdock (2018), since this field most closely resembles the dominant component of NASA's Earth Science Division, rather than the data in the original NSF report, likely comprised of mainly field geologists. While the data presented in Bernard and Cooperdock reflect PhD recipients in 2016, the most recent demographic data on 2019 geoscience PhD recipients appear similar (NSF 2019), underscoring the fact that no progress persists with respect to diversity in the geosciences. The biological sex and race/ethnicity demographics of the atmospheric science population as a proxy for the Earth Science workforce relevant to NASA is shown in Figure 3.8.

The PhD production rate (236/year in atmospheric science) is probably not reliable for estimating the total NASA Earth science workforce. Instead, taking the 25,684 PIs and Co-Is who submitted to NSPIRES with respect to the Earth Science Division from 2014-2020 as shown in Figure 3.6, and postulating that active scientists submit 2 to 3 proposals per year, then a very rough "ballpark" estimate of the number of active Earth scientists involved in proposals submitting to NSPIRES is on the order of 1,400-2,100. For each of the four research fields of the SMD divisions, the "ballpark" estimates of the total number of active researchers in the fields (between ~1,000 and ~3,000) are very rough estimates. A systematic effort to gather the workforce data in all four areas is needed to allow NASA to accurately measure not only the workforce doing research in these areas but to also evaluate the overlap between fields (e.g., scientists studying exoplanets, planetary magnetospheres, or the impact of Earth's magnetosphere on Earth's atmosphere, etc.).

**Finding:** The total size of the U.S. post-PhD workforce in each of the four divisions of SMD, plus the degree of the overlap between divisions, is very poorly quantified.

The demographics of PhD scientists working in the United States are pretty similar across the four disciplines associated with NASA SMD space missions (albeit with the caveats of the limitations of respective workforce surveys described above) (see Figure 3.8). Women comprise 17-38% of the space science workforce, and with respect to race/ethnicity, the workforce is far less diverse. Note that these studies treat gender and race/ethnicity as separate, independent categories while the intersectionality of gender with race/ethnicity tends to compound marginalization. Approximately 75-82% of the space science workforce identify as White, about 8-13% identify as Asian/Asian American (comparable to representation in the overall STEM workforce), the percentage of the workforce identifying as Latinx/Hispanic varies between the disciplines, and Black/African Americans are the least represented in the overall space science workforce. The non-White populations within the space science workforce are clearly small in number and there is considerable uncertainty with respect to demographic data on these populations associated with the range in: methodologies used to collect these data, wording of questions

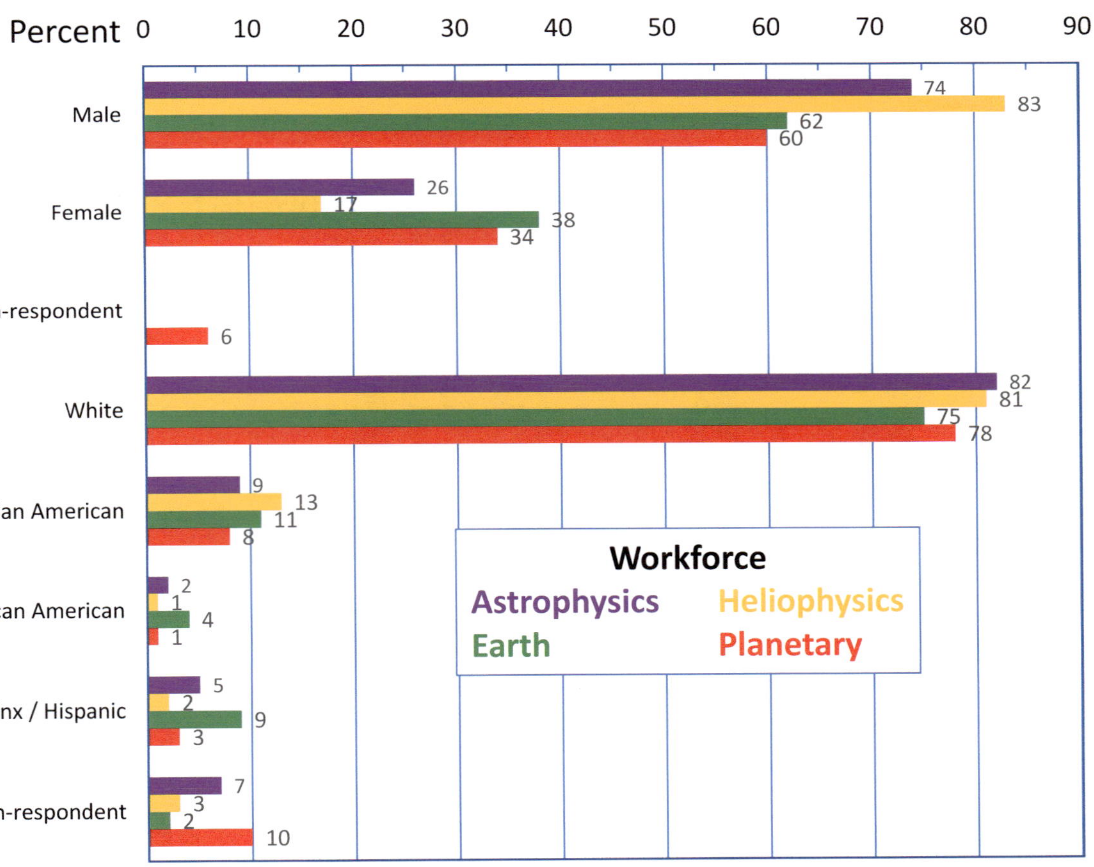


**FIGURE 3.8** Demographics of the space science research workforce (PhD scientists working in the United States) derived from workforce surveys. SOURCES: Based on data from Bernard and Cooperdock (2018), NSF National Center for Science and Engineering Statistics, Pold and Ivie (2018), Porter and Ivie (2019), Porter et al. (2020), and White et al. (2011b).

on survey instrument, how multiple/mixed races/ethnicities are included, etc. A consistent system for gathering workforce data across the space sciences is needed to accurately measure the diversity of the workforce.

## COMPARISON OF DEMOGRAPHICS OF PROPOSERS AND THE WORKFORCE ACROSS THE FOUR SMD DIVISIONS TO THE U.S. POPULATION

In this final section, the demographics data for mission proposers and all NASA proposers are compared with the overall space science workforce data, which comprises the proposer pool. The comparison is also made with the U.S. workforce in STEM fields as well as overall.

### Gender

Women remain underrepresented among the leaders of competed mission proposal submissions, NSPIRES PIs and Co-Is, and within certain sub-areas of the space science workforce relative to their share of the U.S. STEM workforce and the overall national workforce. As shown in Figure 3.9, in 2019, the most recent year for which data are available, women represented 47% of the total U.S. workforce, 39% of the STEM workforce holding PhDs, and 23% of the STEM workforce holding PhDs in a physical science field.[3] The proportion of women in the astrophysics workforce (26%) exceeded that of the STEM workforce with physical science PhDs, but fell below that of the overall PhD-holding STEM workforce and the national workforce. In planetary sciences and Earth sciences, women composed a larger share of the workforce (35% and 38%, respectively) than they did of the total physical science PhD-holding STEM workforce (23%). The proportion of women in the planetary and Earth science workforce nearly equaled that of women in the PhD-holding STEM workforce. However, women remained underrepresented in the planetary and Earth science workforce relative to their share in the overall U.S. workforce.

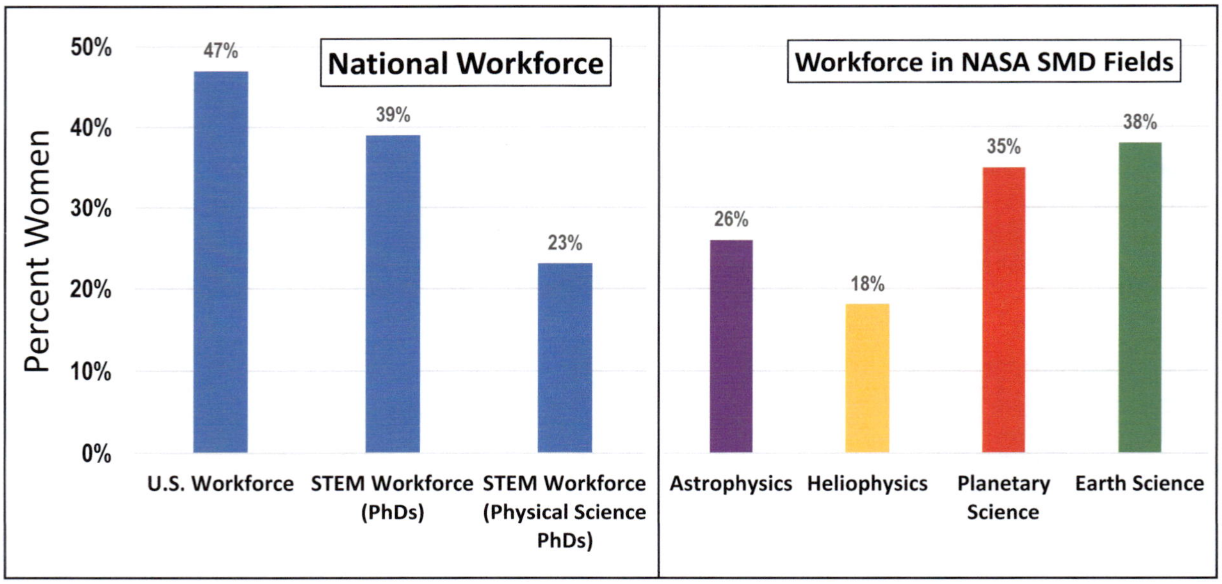


**FIGURE 3.9** Representation of women in the U.S. workforce, the PhD-holding U.S. STEM workforce, and the workforces of the corresponding NASA SMD divisions (Astrophysics, Heliophysics, Planetary Science, and Earth Science). SOURCES: Based on data from Bernard and Cooperdock (2018), NSF National Center for Science and Engineering Statistics, Pold and Ivie (2018), Porter and Ivie (2019), Porter et al. (2020), and White et al. (2011a,b).

[3] See NSF (2021d) and the NSF National Center for Science and Engineering Statistics database "National Survey of College Graduates," https://ncsesdata.nsf.gov/builder/nscg.

The representation of women in heliophysics (18%) remained lower than their share of the STEM workforce (both physical science PhDs and PhDs in any field) and women's share of the U.S. workforce.

Women are also underrepresented among NSPIRES PIs and Co-Is between 2014 and 2020, relative to their share in the U.S. workforce, the STEM workforce and within certain sectors of the space science workforce. As shown in Figure 3.10, 19% of NSPIRES PIs and Co-Is in astrophysics were women, which is lower than women's representation within the astrophysics workforce, the STEM workforce and the U.S. workforce. In planetary sciences and Earth sciences, a similar pattern of underrepresentation is observed. In planetary science, 23% of NSPIRES PIs and Co-Is between 2014 and 2020 were women compared to 35% of the planetary science workforce. In Earth science, 22% of NSPIRES PIs and Co-Is between 2014 and 2020 were women, compared to 38% of the workforce in that SMD science field. In heliophysics, 19% of NSPIRES PIs and Co-Is between 2014 and 2020 were women, which slightly exceeds women's share of the heliophysics workforce.

As previously mentioned, gender is inferred in the demographic data on PIs for competed mission proposals provided to the committee by NASA SMD. Figure 3.10 also shows the inferred gender statistics for competed mission PIs compared with the workforce in NASA SMD fields, the STEM workforce holding PhDs in physical sciences, the overall U.S. STEM PhD workforce, and the overall U.S. workforce. The PIs on the competed mission proposals, both submitted and funded, comprise of 5% to 22% women across the four NASA SMD fields. While this range is comparable to the percentage of women in the heliophysics workforce, it is significantly below the percentages of women in the workforce for the other three divisions.

The percentage of PIs on competed mission proposals that are women vary considerably across divisions. Between 2014 and 2020, very few mission proposals with women PIs were submitted both in astrophysics and Earth science, and none were selected (see Figure 3.2). The percentage of competed mission proposals with women PIs submitted and funded in Heliophysics is comparable (8% versus 9%, respectively; see Figure 3.2). In contrast, 22% of competed missions proposals with women PIs were submitted in planetary sciences, and 40% of the funded missions were those led by women PIs. For comparison, 39% of the U.S. STEM PhD workforce is female which drops to 23% when only the physical sciences are considered.

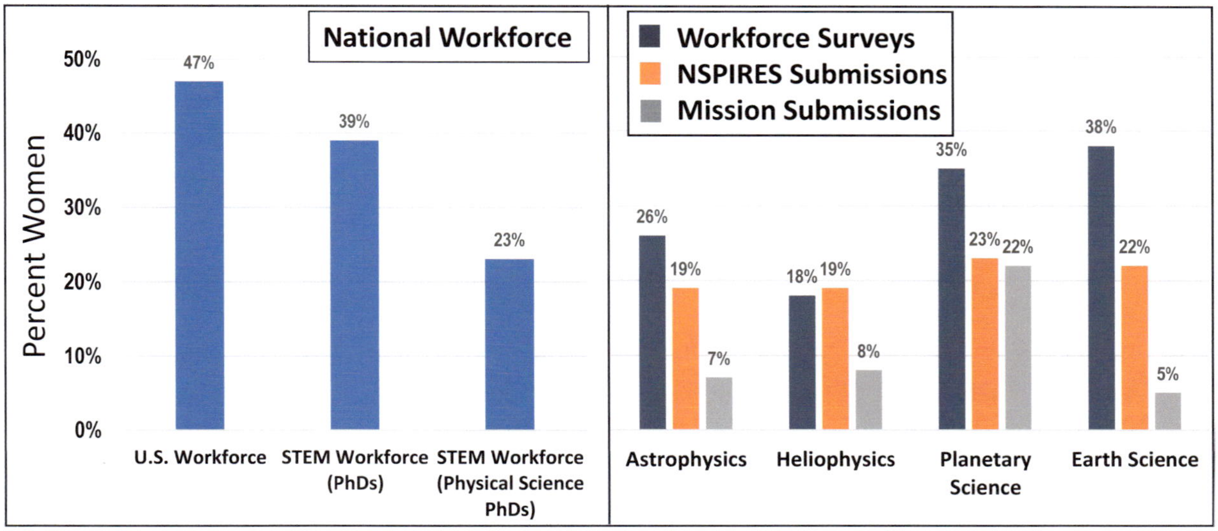


**FIGURE 3.10** Gender statistics of competed mission team leaders compared with the workforce of corresponding NASA SMD divisions and the U.S. STEM workforce overall. Female percentage of populations: PI of submitted mission proposals, PIs and Co-Is of NSPIRES proposals, workforce for SMD science fields, U.S. STEM workforce (Physical Science PhDs), U.S. STEM workforce (PhDs in any field), total U.S. workforce. SOURCES: Based on data from Bernard and Cooperdock (2018), NSF National Center for Science and Engineering Statistics, Pold and Ivie (2018), Porter and Ivie (2019), Porter et al. (2020), and White et al. (2011a,b).

### Race/Ethnicity

Though data on the race/ethnicity of the PIs and Co-Is of competed mission proposals are not available, an examination of data on NSPIRES PIs and Co-Is between 2014-2020 (serving as a proxy for the potential community of mission proposers in the context of this study), and the most recently available space science workforce data, reveal that many racially minoritized populations continue to experience severe underrepresentation relative to their share of the U.S. STEM workforce and the overall U.S. workforce. Note again that these studies treat gender and race/ethnicity as separate, independent categories while the intersectionality of gender with race/ethnicity tend to compound marginalization. As shown in Figure 3.11, 6% of the U.S. workforce identified as Asian/Asian American, compared to 26% of PhD-holding STEM workforce and 29% of the STEM workforce with doctorates in physical sciences. Asian/Asian Americans were less represented in the space science workforce (ranging from 8% in planetary science to 13% in Heliophysics), but were overrepresented within the pool of NSPIRES PIs and Co-Is, relative to their share of the workforce in each of the SMD science fields.

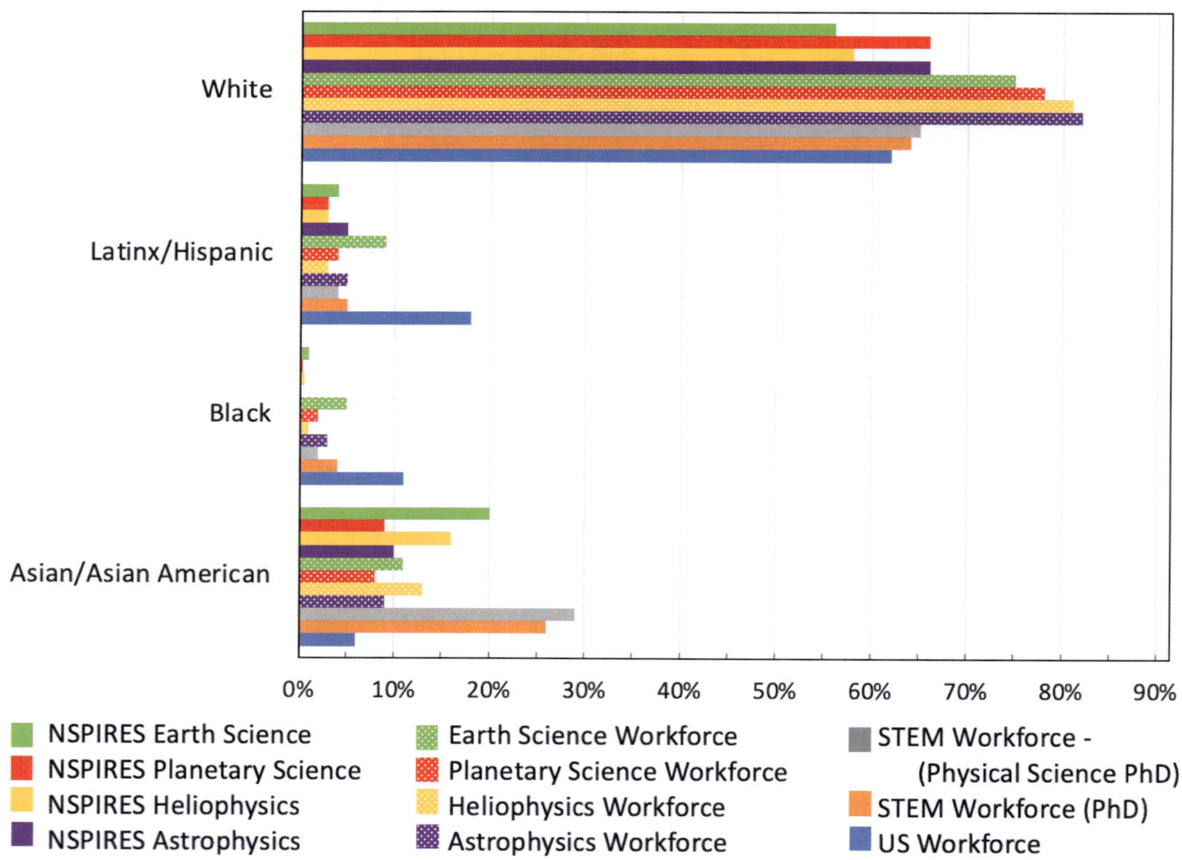


**FIGURE 3.11** Race/ethnicity statistics of NSPIRES PIs/Co-Is compared with the workforce of corresponding NASA SMD division science field and the U.S. STEM workforce overall. The percentages reflect the indicated racial/ethnic group's share of the following populations: PIs and Co-Is of NSPIRES proposals, workforce for SMD science fields, U.S. STEM workforce (physical science PhDs), U.S. STEM workforce (PhDs in any field), total U.S. workforce. Note that the NSPIRES data included about 20% Prefer Not to Answer responses. SOURCES: Based on data from Bernard and Cooperdock (2018), NSF National Center for Science and Engineering Statistics, Pold and Ivie (2018), Porter and Ivie (2019), Porter et al. (2020), and White et al. (2011a,b).

Individuals identifying as Black as well as those identifying as Latinx/Hispanic experience severe underrepresentation in the STEM workforce, the space science workforce, and among NSPIRES PIs and Co-Is, relative to their share of the overall U.S. workforce. Though 11% of the U.S. workforce identify as Black/African American, this population represents just 4% of the doctorate-holding STEM workforce, between 1% and 5% of the space science workforce and less than 1% of NSPIRES PIs and Co-Is between 2014 and 2020. Individuals of Latinx/Hispanic ethnicity are also underrepresented in the STEM workforce, the space science workforce and among NSPIRES PIs and Co-Is relative to their share of the overall U.S. workforce. However, in astrophysics and heliophysics, the proportion of the space science workforce of Hispanic/Latinx ethnicity is similar to that of Hispanic/Latinx NSPIRES PIs and Co-Is. In planetary sciences and Earth sciences, the proportion of the respective workforce identifying as Latinx/Hispanic exceeded this population's share of NSPIRES PIs and Co-Is.

Figure 3.11 also illustrates the overrepresentation of self-identified non-Hispanic White individuals in all four NASA SMD fields relative to their share of the U.S. workforce and STEM workforce. The proportion of NSPIRES PIs and Co-Is identifying as White falls between 56% and 66% in the four NASA SMD fields, which is below their share of the larger space science workforce. However, it is important to note that a significant proportion (18-22%) of NSPIRES PIs and Co-Is decline to disclose their racial/ethnic identity on the demographic questionnaire within the proposal system. Given roughly one-fifth of PIs and Co-Is opt out of the race/ethnicity question, it is difficult to fully characterize the degree to which the racial/ethnic groups are overrepresented or underrepresented among the NSPIRES proposer pool. Additionally, the lack of data on the racial/ethnic identities of competed mission proposal PIs makes it difficult to determine whether the patterns of underrepresentation and overrepresentation within the larger space science workforce and NSPIRES PIs and Co-Is described above hold for the competed mission proposer pool. Nonetheless, given the low representation of Black and Latinx/Hispanic among NSPIRES proposer pool, it is reasonable to assume that these groups experience underrepresentation within the competed space mission proposer pool as well.

While this chapter describes the representational patterns within the space science workforce and among NASA PIs and Co-Is, it is also important to understand the root of these patterns so that points of intervention are identified. The causes of this underrepresentation, which are multi-faceted and operate at multiple points along the education-to-career pathway, are discussed in Chapter 4 of this report.

**Finding:** With the exception of Planetary Science, the percentages of competed mission proposal submissions made by women PIs over 2010-2019 (8% for Heliophysics, 7% for Astrophysics, 22% for Planetary Science, and 5% for Earth Science) are well below the current percentage of PhDs awarded to women in physical sciences (23%), below the percentage of women in the SMD workforce (17% for Heliophysics, 26% for Astrophysics, 34% for Planetary Science, and 38% for Earth Science), and below the percentage of women submitting to the NASA Solicitation and Proposal Integrated Review and Evaluations System (NSPIRES) (19% for Heliophysics, 19% for Astrophysics, 23% for Planetary Science, and 22% for Earth Science).

**Conclusion 3-4:** Further study is needed to evaluate to what extent the differences in the proportion of women among mission PI proposers compared to the proportion of women among doctorate recipients in the physical sciences are due to the mismatch between the career stage distribution of women in these various populations and the current seniority expectation for mission PIs.

**Conclusion 3-5:** Currently, there is no strategy and methodology being employed to effectively gather data on participation in the Earth and space sciences along the whole career pathway—from undergraduate and graduate programs to academic departments to the professional workforce.

## ANNEX 3.A

### Demographics Data Collection in the Earth and Space Sciences

There are wide ranges in quality and quantity of demographics data gathered via the workforce surveys of the different research fields in space sciences as well as via the NASA Solicitation and Proposal Integrated Review and Evaluations System (NSPIRES) system. Table 3.A.1 lists the questions asked by the different methods and the response options.

**NASA's NSPIRES Grant Submission Interface** (NSPIRES).[4] NSPIRES collects demographic data on gender (Note: the question is labeled gender but actually asks about biological sex), ethnicity, race, disability status, etc.[5] The web portal does allow one to click a box "why this information is being collected." However, the language in the box appears to be compliance-driven, and does not explain the importance of collecting such data. Additionally, in its current form, respondents are told that "*Submission of the requested information is voluntary and will not affect an organization's eligibility for an award. However, withholding this information will undermine the usefulness of the information provided by others. Any individual who would prefer not to submit some or all of the information requested should check the box(es) provided.*" Respondents are also required to open additional boxes and pages in order to provide all of the requested information.

**Workforce Survey of 2018 U.S. AAS Members** (Pold and Ivie 2018). The survey was administered by the American Institute of Physics' Statistical Division who sent the survey to the email addresses of American Astronomical Society (AAS) members. The survey of the astronomy workforce included questions about gender identity (including transgender identity), sexual identity, race or ethnicity, disability and accessibility aids. The entrance page to the survey says "Welcome to the AAS Workforce Survey, sponsored by the Demographics Committee. Thank you for your participation!" The demographics questions appear at the very end of the survey, following a range of questions about marital status, children, and childcare.

**2011 Survey of Solar, Space & Upper Atmospheric Physicists** (White et al. 2011a). The survey was administered by the American Institute of Physics's (AIP's) Statistical Division who sent the survey to the email addresses of all members of the Space Physics and Aeronomy section of the American Geophysical Union, the AAS Solar Physics Division, attendees of the Space Weather Week conference, and National Science Foundation principal investigator lists. The survey of the heliophysics community had the least demographics questions (perhaps reflecting it was carried out over a decade ago), one on sexual identity and one on race or ethnicity. These two questions appear at the very end of the survey following open-ended questions about perceived determinants and barriers of respondent's success and preceding a final question requesting year of birth.

**2020 Survey of the Planetary Science Workforce** (Porter et al. 2020). The survey was administered by the AIP's Statistical Division who sent the survey to the email addresses of the Division of Planetary Sciences of the AAS, the Planetary Division of the Geological Society of America, and attendees of the Lunar and Planetary Science Conference. The survey included a series of demographic questions on gender identity, transgender status, sexual identity, racial/ethnic designation, disability and use of accessibility aids. These demographic questions appear toward the end of the survey before open-ended questions on determinants and barriers of respondent's success, and following a range of questions about the use of childcare resources or grants for attending conference meetings. The preamble to the survey says "Thank you for helping us learn more about the demographics and career paths of planetary scientists. Aggregated results will be shared with the community to be used to inform the Planetary Science Decadal Survey which is including workforce issues for the first time."

[4] See the NASA Solicitation and Proposal Integrated Review and Evaluations System at https://nspires.nasaprs.com.

[5] Why this information is being collected: The federal government has a continuing commitment to monitor the operation of its review and award processes to identify any inequities based on gender, race, ethnicity, or disability. NASA asks that you provide information about your gender, race, ethnicity, and disability status to ensure compliance with Title VI of the Civil Rights Act of 1964, 42 U.S.C. § 2000d et seq., Title IX of the Education Amendments of 1972, 20 U.S.C. § 1681 et seq., Section 504 of the Rehabilitation Act of 1973, 29 U.S.C. § 701 et seq., and NASA's implementing regulations at 14 CFR §§ 1250, 1251, and 1253.

**TABLE 3.A.1** Demographics Questions in NSPIRES and Workforce Surveys

| **NASA NSPIRES Person Profile Questions** | |
|---|---|
| ***Question/Category*** | ***Options*** |
| Gender[1] *(choose one)*<br><br>[1]*Note this question is labeled gender but actually asks about biological sex.* | Male<br><br>Female<br><br>Other<br><br>I prefer not to report my gender |
| Ethnicity *(choose one)* | Hispanic or Latin<br><br>Not Hispanic or Latin<br><br>I prefer not to report my ethnicity |
| Race *(select one or more)* | American Indian or Alaskan Native<br><br>Asian<br><br>Black or African American<br><br>Native Hawaiian or Other Pacific Islander<br><br>White<br><br>Other<br><br>I prefer not to report my race |
| Do you have any of the following disabilities/ serious health conditions? Consider your answers without the use of medication and aides (except eyeglasses) or the help of another person.<br><br>*(Check all boxes that apply to you)* | Hearing Impairment<br><br>Visual Impairment<br><br>Mobility/Orthopedic Impairment<br><br>Other<br><br>None<br><br>I prefer not to reveal my disabilities/health conditions |
| **Workforce Survey of 2018 U.S. AAS Members** | |
| ***Question/Category*** | ***Options*** |
| What is your gender identity? | Woman<br><br>Man<br><br>Another identity (please specify if you wish) Prefer not to respond |
| Do you identify as transgender? | Yes<br><br>No<br><br>Prefer not to respond |

*continued*

**TABLE 3.A.1** Continued

| | |
|---|---|
| Do you identify as | Heterosexual or straight<br>Gay or lesbian<br>Bisexual<br>Other<br>Prefer not to respond |
| What is your race or ethnicity *(Please check all that apply)* | American Indian or Alaska Native<br>Asian or Asian American<br>Black or African American<br>Hispanic or Latino<br>Native Hawaiian or other Pacific Islander<br>White<br>Other (please specify)<br>Prefer not to respond |
| Please indicate which of the following apply to you: *(Please check all that apply)* | I am deaf or hard-of-hearing<br>I have difficulty seeing even when wearing glasses<br>I have serious difficulty standing, walking or climbing stairs<br>I have a cognitive or learning disability<br>I have a mental illness<br>I am neuroatypical<br>I have an autoimmune or pain disorder, or other chronic condition<br>I have disabling allergies, asthma, or other environmental sensitivities<br>Other disability:<br>None of the above<br>Prefer not to respond |

**TABLE 3.A.1** Continued

| | |
|---|---|
| Which of the following accessibility aids do you use? *(Please check all that apply)* | Sign language (American or other)<br>Hearing aids, headphones, and other audio devices<br>Closed-captioning<br>Screen readers<br>Braille<br>Speech transcription<br>Mobility aids<br>Note Takers<br>Quiet spaces<br>Dietary accommodation related to health or disability<br>Environmental adjustments (lighting, air, noise, etc.)<br>Service animal<br>Other accessibility aid:<br>None of the above<br>Prefer not to respond |
| Have you requested accessibility accommodation at work or school? *(Please check all that apply)* | Yes<br>No<br>No, but I would benefit from accessibility accommodation<br>No, and I am afraid of disclosing my disability or chronic condition<br>Not applicable |
| **2011 Survey of Solar, Space & Upper Atmospheric Physicists** | |
| ***Question/Category*** | ***Options*** |
| Are you: | Female<br>Male |
| To what racial or ethnic group do you belong? (Check all that apply.) | Asian or Asian American<br>Black or African American<br>Hispanic or Latino<br>White<br>Other |

*continued*

**TABLE 3.A.1** Continued

| **2020 Survey of the Planetary Science Workforce** | |
|---|---|
| ***Question/Category*** | ***Options*** |
| What is your gender identity? | Woman<br>Man<br>Another identity (please specify if you wish)<br>Prefer not to answer |
| Do you identify as transgender? | Yes<br>No<br>Prefer not to answer |
| Do you identify as: | Heterosexual<br>Gay or Lesbian<br>Bisexual<br>Other (please specify if you wish)<br>Prefer not to answer |
| Which of the following racial/ethnic designations best describes you? *Please select all that apply.* | Hispanic/Latinx<br>American Indian or Alaska Native<br>Asian or Asian American<br>Black or African American<br>Native Hawaiian or Other Pacific Islander<br>White<br>Addition (please specify if you wish)<br>Prefer not to answer |
| Please indicate which of the following apply to you. *Please select all that apply.* | I am deaf or hard-of-hearing<br>I have difficulty seeing even when wearing glasses<br>I have serious difficulty standing, walking, or climbing stairs<br>I have a cognitive or learning disability<br>I have a mental illness<br>I am neuroatypical<br>I have an autoimmune or pain disorder, or other chronic condition<br>I have disabling allergies, asthma, or other environmental sensitivities<br>Other (please specify if you wish)<br>None of the above<br>Prefer not to answer |

**TABLE 3.A.1** Continued

| | |
|---|---|
| Which of the following accessibility aids to you use?<br><br>*Please select all that apply.* | Sign language, American or other<br><br>Hearing aids, headphones, and other audio devices<br><br>Closed-captioning<br><br>Screen readers<br><br>Braille<br><br>Speech transcription<br><br>Mobility aids<br><br>Note takers<br><br>Quiet spaces<br><br>Dietary accommodation<br><br>Environmental adjustments – For example lighting, air, noise, etc. Service animal<br><br>Other accessibility aid (please specify)<br><br>None of the above<br><br>Prefer not to respond |
| Have you requested or utilized accessibility accommodations at work or school? *Please select all that apply.* | Yes<br><br>Yes, but accommodations were not available<br><br>No<br><br>No, but I would benefit from accessibility accommodations<br><br>No, and I am afraid of disclosing my disability or chronic condition<br><br>Not applicable |

SOURCE: Data from the NASA Solicitation and Proposal Integrated Review and Evaluations System (https://nspires.nasaprs.com/external), Pold and Ivie (2018), and White et al. (2011b).

# 4

# Pathways into Space Sciences

At any given time, the pool of potential National Aeronautics and Space Administration (NASA) mission leaders represents the culmination of a long series of educational and research training experiences through which they acquired the knowledge and skills appropriate to the rigor and challenge of mission leadership. This set of knowledge and skills ranges from deep (usually PhD-level) scientific expertise in a relevant subject area, to hands-on (oftentimes engineering-related) experience with instruments and/or data, to proposal development experience (ideally directly), to management experience at one or more leadership levels within the typical hierarchy of NASA space missions.

Acquiring such knowledge and experiences takes time and begins with entry into higher education, where most scientists and engineers first begin to acquire foundational knowledge as well as exposure and involvement in authentic research. The committee defines and discusses what constitutes authentic research experiences below, but to put it simply, it means "real" research or experiences that involve direct engagement with the actual work of missions or that provides training that is directly relevant to real mission work.

If a scientist or engineer first becomes eligible to help lead a NASA space mission when they secure professional employment in the field, then this pathway spans *at least 20 years* and transitions across multiple key junctures. This generally includes starting college to graduation (~4 years), starting graduate school to doctoral degree (~6 years), postdoctoral research training to professional position (3-5 years), and becoming an established professional (~7 years). Today's emerging leaders who may just now be proposing a first NASA mission were beginning their journey as college freshmen around the start of the millennium.

## DEMOGRAPHICS ALONG THE PATHWAY: ATTRITION OF TALENT AT THE SOURCE

Reasonably good statistics are available from comprehensive national databases for the first half of this journey, from first-year undergraduate to PhD. For example, Table 4.1 depicts a synthetic cohort analysis of this pathway for U.S. citizens and permanent residents in physics and astronomy, as entering first-year students in 2007 to PhD in 2018. Physics and astronomy are linked by the fact that students who eventually earn PhDs in astronomy and astrophysics often begin as physics majors. This is a "synthetic cohort" in the sense that it does not represent a literal longitudinal tracking of the same individuals over time; rather, the experience of the cohort is inferred by comparing national demographics data at time points separated by the typical duration of various academic stages. There are also limitations to a simple, linear "pipeline" progression model; however, it does provide a convenient basis for useful comparisons. The data are meant to be illustrative of major representative areas relevant to NASA Science Mission Directorate (SMD) space missions, and reveal important patterns of major missed opportunities and ongoing disparities into the ranks of potential future NASA mission leaders. First, only about 2% of all first-year college students in the United States express an interest in a physical sciences major. Encouragingly, this (relatively small) cohort of incoming, interested students does not exhibit large disparities by race/ethnicity (on the other hand, women in the aggregate do experience disproportionate attrition in physics during the transition from high school to college; see Figure 4.3 below). However, of these initially interested students, only about 11% of White students complete a physics or astronomy bachelor's degree (consistent with an analysis from University of California, Davis (see Figure 4.1); and only 4% of underrepresented minority (URM) students[1] with similar interests do so, a disparity of about a factor of 3. While the largest attrition appears to occur early (i.e., prior to formally declaring the major) in the undergraduate experience, significant losses continue, with roughly 25% of White students switching out after declaring their major and roughly 40% of URM students doing so.[2] While there is no longer a significant ethnic/racial disparity (aggregated across URMs) between the baccalaureate and PhD stages (the combination of ~30% graduate admission rate and ~60% PhD completion rate for those admitted are similar for all groups, see Table 4.1), there is a large disparity at the undergraduate level. This culminates in a very low number of URM PhDs in physics and astronomy. This has obvious long-term consequences for the diversity of the profession at the postdoctoral level and beyond.

**TABLE 4.1** Physics and Astronomy Synthetic Cohort from College First Year to PhD

| | U.S. Citizens[a] | White | URM[b] |
|---|---|---|---|
| First year, first-time undergraduates (2007)[c] | 2,764,690 | 1,655,714 | 766,844 |
| Estimated number intending physical sci major (2007)[d] | 66,000 | 41,000 | 11,500 |
| *... of whom, __% complete physics or astronomy degrees* | *10%* | *11%* | *4%* |
| Bachelor's degrees in physics and astronomy (2012)[e] | 6,664 | 4,596 | 473 |
| *... of whom, __% are admitted to graduate programs* | *29%* | *Not known* | *Not known* |
| Entering grad program[f] in physics/astronomy (2012)[e] | 1,937 | *Not known* | *Not known* |
| *... of whom, __% complete the PhD in 6 years* | *59%* | *Not known* | *Not known* |
| PhD in physics and astronomy (2018)[e] | 1,151 | 805 | 76 |
| *Overall retention from bachelor's to PhD* | *17%* | *18%* | *16%* |

[a] Includes citizens and permanent residents in the categories at right (White, URM) and those identifying as "other," as well as racial/ethnic minority groups defined by the federal government as over-represented in STEM fields relative to their proportion in the population of U.S. citizens and permanent residents (e.g., Asian American).

[b] URM (underrepresented minority) refers to the racial/ethnic groups defined by the federal government as minorities who are underrepresented in STEM fields relative to their proportion in the population of U.S. citizens and permanent residents. URM groups so defined currently include African Americans, Hispanic Americans, and Native Americans including Alaska Natives and Native Hawaiians.

[c] Enrollment of first-time, first-year undergraduate students at all institutions, by citizenship, ethnicity, race, sex, and enrollment status, Table 2.2, 2004-14 (2013 Women, Minorities, and Persons with Disabilities in Science and Engineering: 2017. Special Report NSF 17-310. Arlington, VA., WMPD, www.nsf.gov/statistics/wmpd).

[d] Based on numbers in Appendix Table 2.16 Freshmen intending science and engineering (S&E) major by, by field, sex, and race or ethnicity, 1998-2012, [NSF Science and Engineering Indicators, 2016]. Unfortunately, the number of entering first-year students who intend to major specifically in physics or astronomy is not known.

[e] AIP Enrollments and Degrees Survey.

[f] Includes MS and PhD students.

---

[1] See Table 4.1 for definition.

[2] See Chapter 2 of *Talking About Leaving Revisited* (Thiry et al. 2019).

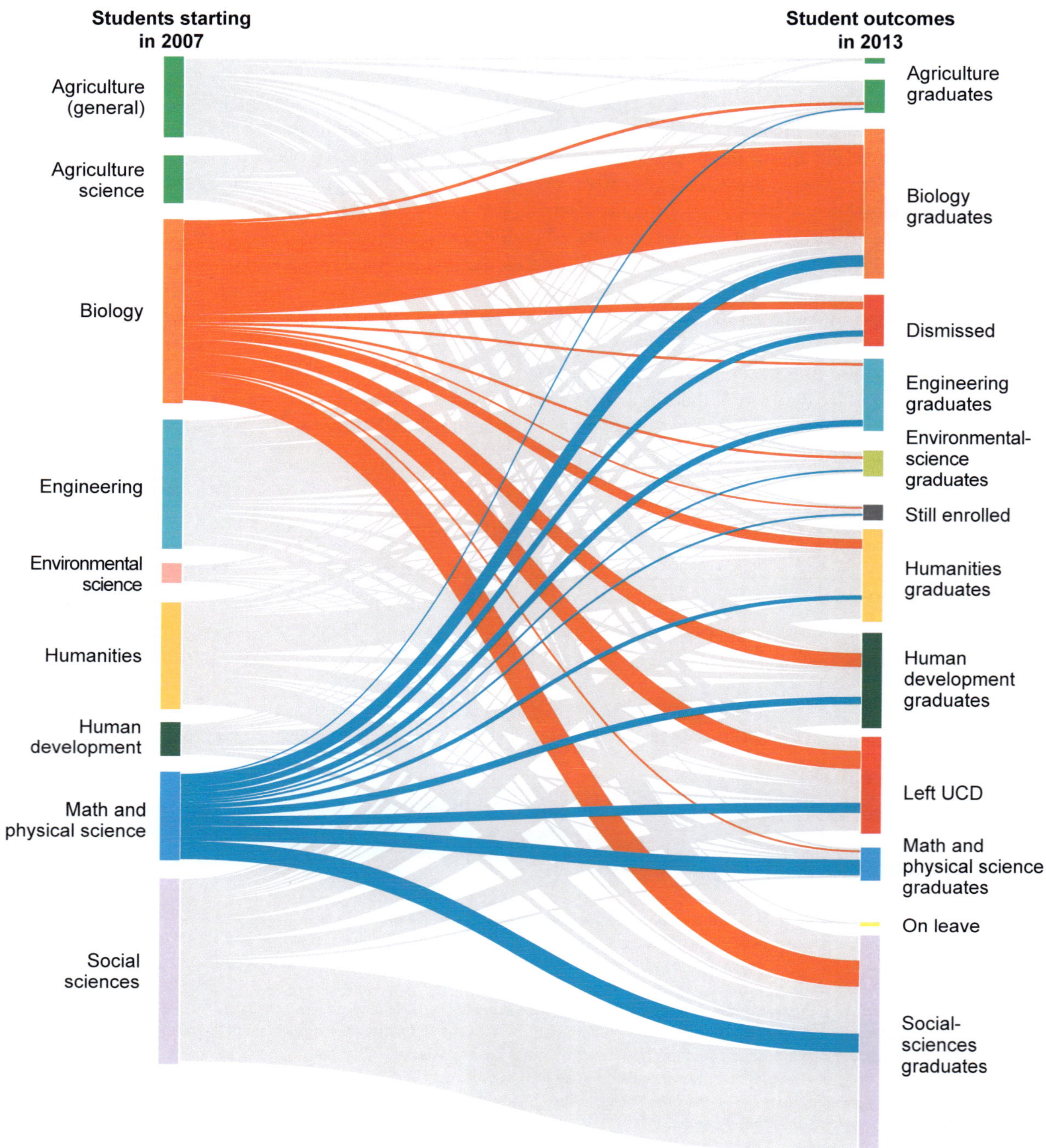


**FIGURE 4.1** Results from a study of undergraduate degree outcomes versus incoming student interests, by field. In contrast to other STEM disciplines like biology or social sciences, physical sciences lose ~90% of students arriving at college with an interest in those fields. SOURCE: Reproduced from M. Molinaro, Center for Educational Effectiveness, University of California, Davis, as published in Bradforth, S. E., E. R. Miller, W. R. Dichtel, A. K. Leibovich, A. L. Feig, J. D. Martin, K. S. Bjorkman, Z. D. Schultz, and T. L. Smith. 2015. University learning: Improve undergraduate science education. *Nature* 523:282-284.

Indeed, the severity of, and the disparity at, the undergraduate "pinch point" results in a very small number of PhDs from minoritized communities in the disciplines most relevant to NASA mission leadership. For example, the number of African Americans and Hispanic Americans combined receiving PhDs in astronomy in any given year has never exceeded single digits, and the number of African American PhDs in astronomy in any given year over the past decade has averaged ~1 (see Figure 4.2). In the case of Native Americans, a total of 3 Native Hawaiians have received a PhD in astronomy over the past 50+ years of available statistics (data not shown) (NASEM 2021).

**Finding:** The number of African Americans, Hispanic Americans, and Native Americans combined receiving PhDs in astronomy in any given year has never exceeded single digits, the number of African American PhDs in astronomy in any given year over the past decade has averaged ~1, and the number of Native American PhDs in astronomy in any given year remain too small to count.

**Conclusion 4-1:** Significant and concerted efforts may be needed to ensure that the currently small pool of scientists of color have every opportunity to engage in NASA mission-related work and leadership.

Interestingly, while women continue to be highly underrepresented in physics and astronomy at all levels, there does not currently appear to be similar attrition from undergraduate to PhD levels, and even at the early-career faculty level, though a significant drop-off in representation still appears at senior ranks (see Figure 4.3). In addition, other physical science fields like chemistry demonstrate that representation at or near parity for women is possible all the way to the PhD level (see Figure 4.3 and NSF [2021d]). These patterns are consistent with longstanding gendered and racialized patterns of persistence and attrition in science, technology, engineering, and mathematics (STEM) in general and the physical sciences specifically (Thiry et al. 2019; NSF 2021d). As presented in the aggregate, it is also important to note that these patterns do not reflect potential for intersectional variation—for example, gender differences among Black or Latinx students or racial differences among women.

**Finding:** As the main disciplinary pathways to future NASA mission leadership, the physical sciences currently have a very small "market share" of college students (~2%). Only about 11% of White and Asian American students, and only about 4% of URM students, in the physical sciences persist to the baccalaureate degree in physics/astronomy.

**Conclusion 4-2:** The very low overall retention (~11%) in NASA SMD relevant disciplines during undergraduate training, and the accompanying racial/ethnic disparity (factor of ~3), is currently a major "pinch point" that restricts the size and diversity of the pool of PhD scientists for future NASA mission leadership.

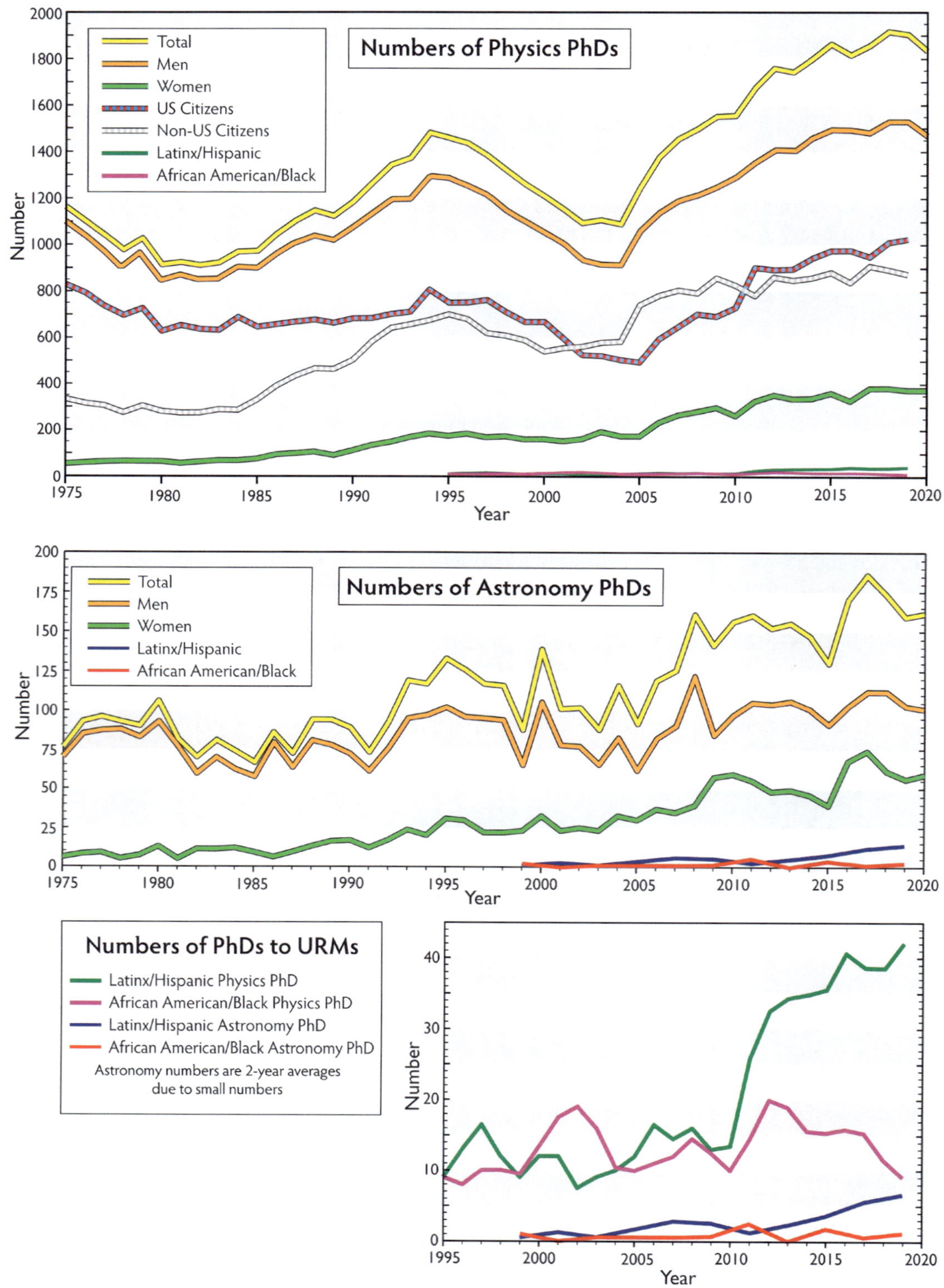


**FIGURE 4.2** Trends in physics and astronomy PhDs: (*Top*) Physics PhDs disaggregated by gender and citizenship status as well as race/ethnicity; (*Middle*) Astronomy PhDs disaggregated by race/ethnicity and gender. (*Bottom*) Black/African American and Hispanic/Latino American PhDs in physics and astronomy. SOURCE: Produced using data from the Statistical Research Center of the American Institute of Physics, see https://www.aip.org/statistics.

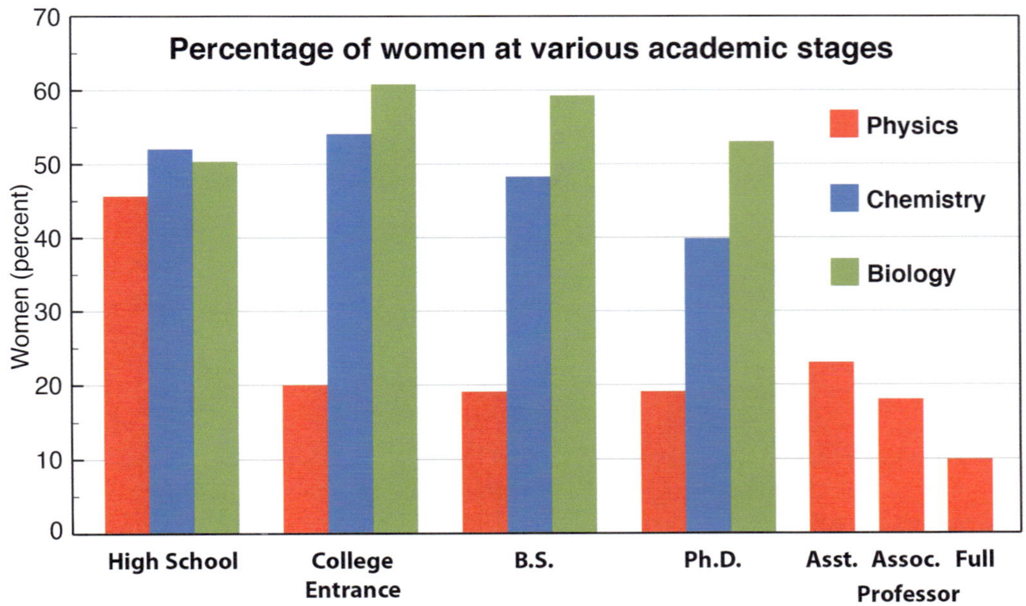


**FIGURE 4.3** Percentage of students and faculty in physics, chemistry, and biology who are women at various academic stages in 2018. SOURCES: Based on data from Hazari (2018) and Porter and Ivie (2019).

## THE ROLE OF AUTHENTIC RESEARCH EXPERIENCES

The fact that (a) such a large rate of attrition occurs among people who initially express a desire to pursue careers in NASA SMD-related disciplines, and (b) the largest point of attrition occurs early in the undergraduate experience, has been recognized by researchers for decades (NAS et al. 2007; Beasley 2011; NASEM 2020a). Seminal ethnographic studies such as *Talking About Leaving* (Seymour and Hewitt 1997), *Talking About Leaving Revisited* (Thiry et al. 2019), and *Opting Out* (Beasley 2011), along with four decades of research on these processes have documented the confluence of negative experiences that drive otherwise talented and capable students away from the science and engineering (S&E) fields. These reasons include, but are not limited to:

1. The lack of opportunity for early and ongoing real research experiences (i.e., an emphasis on experiences with *authentic research* or "actually doing" science over pedantic "gateway" courses;[3] see the National Academies report *Undergraduate Research Experiences for STEM Students* (NASEM 2017b) and lack of resources in schools and universities to involve students in such research;
2. The culture and environment of S&E fields (i.e., a "weed out" mentality that eschews developing talent and instead rewards the few perceived to have "innate brilliance");
3. Structural racism, sexism and interpersonal attitudes and stereotypes (e.g., implicit biases) that deny marginalized and minoritized students access to essential training experiences, networks, etc.; and
4. A desire to pursue careers that are perceived as having a community orientation (e.g., in contrast to addressing abstract questions of physics that are seen as divorced from or even as taking resources away from immediate and pressing health and safety needs of marginalized communities).

[3] The Howard Hughes Medical Institute Science Education program defines authentic research as experiences in which students: formulate and test a hypothesis to make a discovery; perform observational, descriptive, or computer-simulated research to interpret, organize, or analyze existing information in new ways; choose a topic, identify resources, and gather, analyze, and share their data; use creativity and inquiry to generate knowledge (not simply replicate or report what is already known); use the techniques of real research done in laboratories and/or in the field; and take ownership of their projects. In the context of this report, we take authentic research experiences to be those that prepare students for future NASA mission roles through involvement with real NASA mission-related work.

**Finding:** Research suggests that the major reasons for such large losses of URM talent along academic pathways include (1) the lack of opportunity for early and ongoing real research experiences (i.e., experiences with authentic research or "actually doing" science); (2) the culture and environment of STEM fields (e.g., a "weed out" mentality); (3) structural racism, sexism and interpersonal attitudes and stereotypes (e.g., implicit biases) that deny marginalized and minoritized students access to essential training experiences and networks; and (4) a perception that physics and astronomy are disconnected from the needs of the communities with whom students identify.

**Finding:** Decades of educational research suggest that early and ongoing experiences with authentic research—experiences that engage students not only in learning about but actually doing research—is key to retaining students generally and URM students specifically. In the context of this report, we take authentic research experiences to be those that prepare students for future NASA mission roles through involvement with real NASA mission-related work.

Not all of these factors can be addressed by NASA SMD, at least not directly. However, NASA SMD does have a significant opportunity to directly and uniquely impact the need for early and ongoing experiences with authentic research. Indeed, in the context of providing individuals with the sorts of research experiences that will prepare them specifically for future NASA mission roles. There is arguably no better preparation than experiences linked to actual NASA missions. Incorporating these experiences and opportunities into missions is best done by the mission teams themselves. This integration requires resources to implement as well as professional support (e.g., from NASA agency-wide efforts such as the Science Activation program). One key area for which mission leaders may need additional professional development and support is in the science of effective mentoring practices. The recent National Academies report *The Science of Effective Mentorship in STEMM*—and its associated online guide with tools for both mentors and mentees—can be leveraged as a resource for this purpose (NASEM 2019). As noted in that report, the demographic realities of pathways into STEM careers means that mentor-mentee relationships will likely involve individuals who do not share gender, racial, and ethnic identities. Such "non-homophilic" mentoring relationships can—and need to be—successful. Mentorship education for both mentors and mentees can help to ensure success. The National Academies report on effective mentorship also includes evidence-based curricula for mentors to enhance the effectiveness and cultural responsiveness of mentoring relationships, in particular to support the development of individuals from historically underrepresented groups in STEM.

In addition, ensuring that talent development is incorporated into mission design at the proposal stage will require that mission teams understand this to be clearly an expectation and thus a component of mission evaluation. *Put plainly, a mission proposal would not be deemed as Excellent overall if its plan for engaging and developing diverse talent for future missions is short of Excellent.*

**Conclusion 4-3:** NASA is uniquely positioned to provide individuals with the sorts of research experiences that are relevant to future mission roles, and there is arguably nothing more authentic than experiences linked to actual NASA missions.

## THE VITAL ROLE OF PARTNERSHIPS WITH MINORITY SERVING INSTITUTIONS

The "pinch point" problem that appears to be endemic to the early undergraduate experience in the physical sciences (see Figure 4.1 above) is not in fact a universal feature of higher-education institutions. Many Minority Serving Institutions (MSIs)—including Historically Black Colleges and Universities (HBCUs), Hispanic Serving Institutions (HSIs), and Tribal Colleges and Universities (TCUs)—have achieved much better outcomes in retention and advancement of students of color generally, and especially in STEM disciplines. For example, according to a report from the American Association of University Professors (AAUP) (Gasman et al. 2007), HBCUs make up just 3% of all institutions of higher education, yet (at the time of the AAUP report in 1996) the top ten producers of African American baccalaureates in the physical sciences are all HBCUs. More recent statistics (Diverse 2021) show that a few large, non-HBCU institutions have entered the top ten, but HBCUs still dominate the list. Moreover, of the top 15 colleges and universities that graduate African Americans who go on to earn PhDs in

STEM fields, 11 are HBCUs. Similarly, of the top 15 colleges and universities that graduate Hispanics or Latinos who go on to earn PhDs in STEM, 9 are HSIs (NAS et al. 2011; AIR 2014).

**Finding:** MSIs—including HBCUs, HSIs, and TCUs—represent a potentially large and diverse talent pool for future NASA missions. For example, HBCUs are consistently the majority of the top ten producers of African American baccalaureates in physics.

And yet historically, MSIs have received only a small portion of the federal R&D funding allocated to institutions of higher learning.[4] The legacy of the disproportionate funding between predominantly White institutions (PWIs) and HBCUs can be seen in the current disparities in STEM facilities including laboratories and capacity for sustaining major programs of research at HBCU campuses. Not a single HBCU is a Carnegie R-1 institution, and therefore have not been targeted for the largest research investments, perpetuating a cycle of under-investment in research capacity. As a result, many HBCU campuses partner with PWIs on sizeable research endeavors and through dual-degree programs; this has implications for whether HBCUs are seen as strong candidates to lead a mission as the primary institution, unless institutional partnerships can be appropriately recognized as strengths in mission competitions (Newman and Jackson 2013).

More than 20 years ago, NASA SMD recognized the value of partnership with MSIs specifically in the context of NASA mission engagement and talent development, and also recognized the need for capacity building at many of these under-resourced institutions. NASA SMD created a special funding program called Minority University and College Education and Research Partnership Initiative (MUCERPI) (Sakimoto and Rosenthal 2005) that supported mutually beneficial partnerships between MSIs and major research institutions. The MUCERPI program operated from approximately 2000-2006, and though it existed for only a short time, some of the most successful and well-known programs of the past 20 years—such as the Fisk-Vanderbilt Masters-to-PhD Bridge Program (Stassun et al. 2011), the Columbia Bridge to the Doctorate Program,[5] the CalBridge Program (Rudolph 2019), and others—got their funding start through MUCERPI.

These exemplar programs have in most cases been "bridge" type programs, through which minoritized students at the undergraduate level are trained in NASA mission relevant research and supported continuously across the transition into graduate-level training, through close partnerships with the MSIs that are the baccalaureate origins of so many of these students (see above). Importantly, these programs continue to be an integral part of the pathway to successful PhD access, especially for minoritized groups, as the programs intentionally capture a population of students who might otherwise be lost at the baccalaureate-to-PhD juncture. As noted in Table 4.1, ethnic/racial disparities (aggregated across URMs) from the baccalaureate to the PhD stages of education and training in physics and astronomy are no longer significant, which is a significant accomplishment in itself and argues for ongoing investment in these vital programs.

**Finding:** A previous NASA SMD-funded program, MUCERPI, focused on training in state-of-the-art research methods and preparation for future NASA leadership specifically through partnerships with minority-serving institutions, has been defunded.

Relatedly, it may also be worthwhile for NASA to consider the equivalent of NSFs EPSCOR kind of program (Melkers and Wu 2009; Wu 2010; Feldman et al. 2014). Indeed, for NASA to have impact on demographic and pipeline issues in mission leadership on decade timescales, investment may be needed at the state level with regards to building up institutional capacity. Thus, it may be beneficial to target funding specifically for areas of the country that could benefit from NASA investment in research to overcome the systemic biases emerging from well-established universities receiving a disproportionate majority of mission funding. To be sure, the recommended

[4] See, for example, Slaton (2010). For instance, during a time of legal separate but equal policies, the Maryland system president Harry Clifton Byrd invested $4 million in infrastructure at UMD College Park and only $100,000 at the HBCU UMD Eastern Shore. From 1940-1949 UMD College Park had an operating budget of $50 million and Eastern Shore had a budget of $500,000. This systematic denial of resources severely limited opportunities for Black participation in STEM.

[5] Columbia University, "Bridge to the PH.D. Program in STEM," https://bridgetophd.facultydiversity.columbia.edu.

NASA partnerships with MSIs can go a long way to address this; the geographic distribution of MSIs and of EPSCOR states overlap substantially.

## THE ROLE OF NASA AGENCY-WIDE EFFORTS

NASA coordinates, at several levels within the organization, a number of diversity and inclusion efforts aimed at improving preparation and early access to research. At the top-most level, NASA has recently launched Mission Equity, an agency-wide assessment of all things NASA does, from procurements and grants, to programs and policies, to examine the obstacles to participation and partnership faced by underrepresented and underserved communities. From within SMD, Science Activation (SciAct), partners with educators to improve U.S. science literacy and advance national education goals. SciAct interacts with educators in every state, supporting collaborative efforts aimed largely at K-12 and informal education, through a broad and congruent ecosystem of activities that align with the federal Co-STEM efforts.

In response to a programmatic review by the National Academies' Board on Science Education (NASEM 2020b), SciAct has further focused their top-level goals through mid-level objectives, including better leveraging of "enhancing connections to" the NASA SMD assets (i.e., those missions the Directorate administers, the research scientists who develop and use those facilities, and the scientific data that is produced). Those objectives also include more directed efforts to reach diverse communities. The National Academies report also recognized that moving away from "the 1% model," while reducing some redundancy, limits access to some of those SMD assets, and that SciAct should look to further engage SMD missions and their scientists in these early education and preparation efforts (NASEM 2020b).

**Finding:** SMD has the unique assets of access to missions, scientists, and data that can be applied to enhancing education and early research experiences, and improving preparation.

The Office of STEM Engagement (OSTEM) strives to enhance K-12 and postsecondary education and access to NASA-wide projects through a centralized office. OSTEM has a specific focus of strengthening support for underrepresented communities, addressing Presidential Executive orders on promoting equal opportunity to URMs,[6] Asian and Pacific Islanders,[7] for participation in federally sponsored programs. NASA Minority University Research and Education Programs (MUREP), which is built off of the successes of the MUCERPI program, provides competitive awards to HBCUs and other MSIs for research involving NASA projects. The MUREP Institutional Research Opportunity (MIRO) further builds upon these opportunities, by increasing institutional competitiveness through infrastructure advancement, analogous to the NSF Mid-scale RI-2. These pan-agency efforts can be leveraged to help institutional partners and NASA itself address some of the attitudinal, cultural, and messaging issues.

**Conclusion 4-4:** MUREP and MIRO through NASA's OSTEM provide specific engagement to underrepresented and underserved communities in STEM, and develop the research capacity and infrastructure of MSIs in areas of strategic importance and value to NASA's mission and national priorities. Further partnership among NASA SMD, OSTEM and MSIs, and leveraging NASA's unique assets, would strengthen URM participation in missions.

[6] These include Executive Order 13532, "Historically Black Colleges and Universities," Executive Order 13555, "White House Initiative on Educational Excellence for Hispanics," and Executive Order 13592, "Improving American Indian and Alaska Native Educational Opportunities and Strengthening Tribal Colleges and Universities."

[7] Executive Order 13515, "Asian American and Pacific Islander Community."

## THE STEM RESEARCH PATHWAYS IN THE NASA MISSION LEADERSHIP CONTEXT

The pathways that students take through the higher-education experience discussed above and then into professional careers can be represented through a braided river schematic such as developed for an American Geophysical Union (AGU) Eos article on the STEM workforce (Batchelor et al. 2021). Figure 4.4 captures this concept for the STEM workforce with proximity to NASA spaceflight missions, with different entry and exit points along the way into the stream of NASA mission experience or principal investigator (PI) status. The figure also attempts to illustrate the multiple opportunities as well as barriers to entry into the river of NASA mission experience. Notably, not all individuals coming into the workforce from the top of the schematic will end up in the PI reservoir in the bottom; they may decide (or be pressured) to leave the field, or they may wind up in one of a myriad other scientific or management roles. However, all researchers that become PIs will need to flow through one of the pathways visualized here.

It is important to note that mission leadership includes other critical team members in addition to PIs, such as Co-Is, deputy PIs, and project scientists, who all play key roles in shaping the science of the mission. Some of these roles can serve as critical stepping stones on the way to becoming a PI. At the same time, some individuals may never aspire to become a PI, preferring the science aspect of missions over management; yet they still play important leadership roles. In fact, the space science community is beginning to more fully recognize the complexities of how science is organized and how teams are rewarded. Indeed, broader recognition of the various kinds of leadership roles in missions suggests that monitoring and assessing mission team diversity is as important as tracking PI demographics.

At least part of the drop in the percentage of women at senior career levels in physics illustrated in Figure 4.3 can be explained by the variation over time in the percentage of PhDs awarded to women in physics and astronomy (Ivie and Nies 2003) (see Figure 3.2). There are clearly structural and cultural factors that encourage some people and discourage others along the career pathways, as reported in the open-ended questions in workforce surveys (White et al. 2011b; Pold and Ivie 2018) and echo decades of research into persistence in STEM.[8] The traditions and cultural norms of the four SMD divisions are different and. recent National Academies decadal surveys,[9] acknowledging these differences, include working groups for addressing the state of the profession for a specific SMD division and making recommendations for improving the climate and diversifying the corresponding workforce. Additionally, a significant number of white papers[10] submitted for decadal surveys describe a diversity of issues faced by women of all races/ethnicities and men from underrepresented groups at different career stages. Chapter 5 of this report includes more discussion of specific barriers and opportunities for space scientists describes along pathways to mission PI-ship as illustrated in Figure 4.4.

---

[8] See Chapter 5 for detailed review of structural, organizational, interpersonal, and intra-personal barriers. Also see NSF (2021d) and Thiry et al. (2019).

[9] NASEM, "Planetary Science and Astrobiology Decadal Survey 2023-2032," https://www.nationalacademies.org/our-work/planetary-science-and-astrobiology-decadal-survey-2023-2032.

[10] Published white papers for Astro2020 Decadal on Astronomy and Astrophysics are available at https://baas.aas.org/astro2020-science, and submitted white papers for Planetary and Astrobiology Decadal 2023-2032 are available at https://www.nationalacademies.org/our-work/planetary-science-and-astrobiology-decadal-survey-2023-2032.

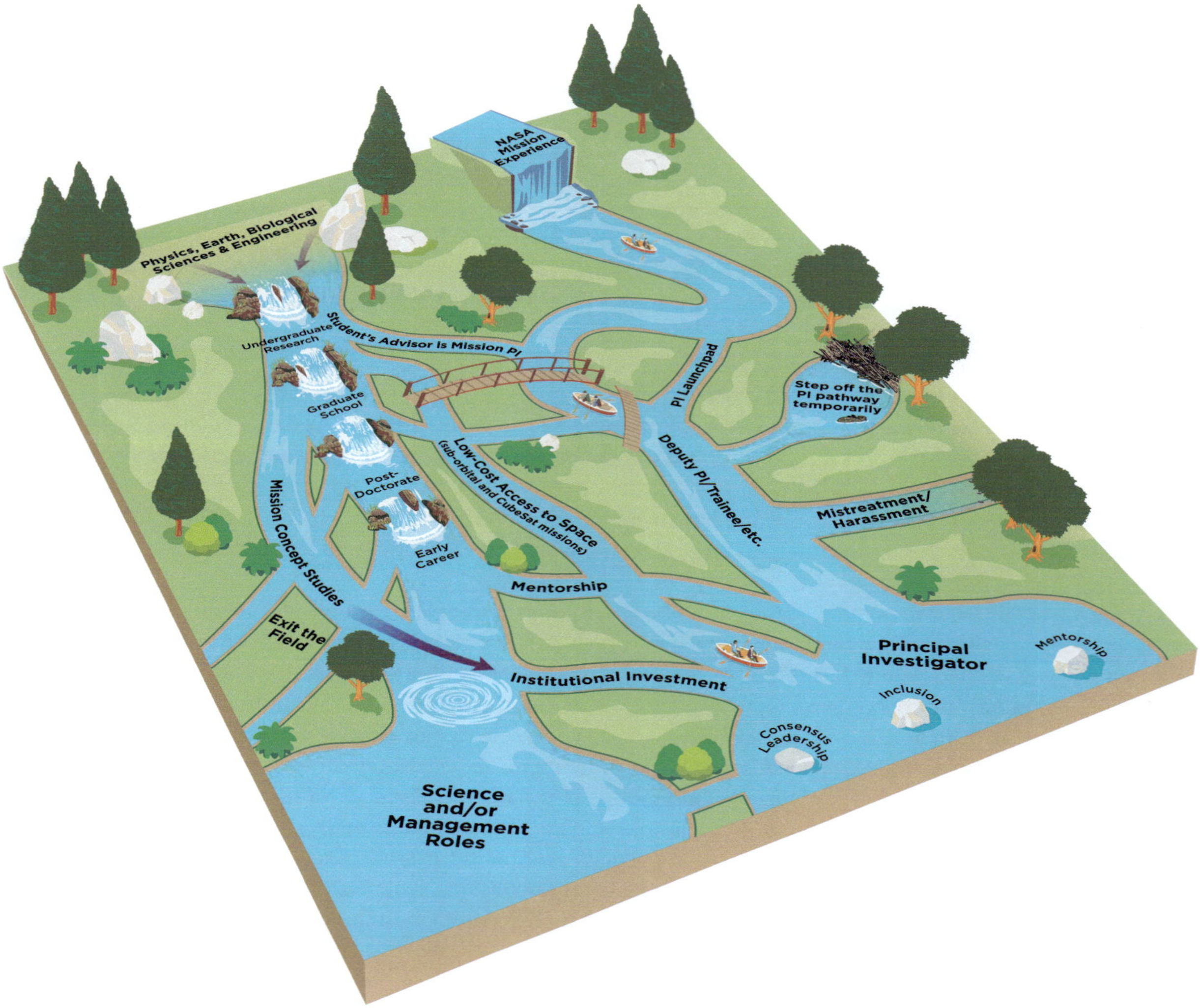


**FIGURE 4.4** Schematic of the science, technology, engineering, and mathematics workforce in proximity to NASA mission experience following the "braided river" concept. The pathways to principal investigator are illustrated, with multiple entry and exit points along the way.

# 5

# Barriers and Opportunities in NASA Space Missions

To increase diversity in competed space missions, it is important to understand the barriers that hinder the development, preparation, and submissions of successful proposals. Also, how these barriers can have a disproportionate impact on populations that remain underrepresented in the space science workforce (e.g., women of all racial/ethnic identities, racially minoritized scientists of all gender identities, gender and sexual minorities, individuals with disabilities, etc.). Drawing on the research literature on advancing diversity in the scientific workforce and enterprise as well as the National Opinion Research Center (NORC) at the University of Chicago qualitative research study[1] on the experiences of competed space mission principal investigators (PIs), this chapter outlines the psychosocial and large-scale factors that act as barriers to increasing the diversity among competed space mission leadership. The recommendations outlined in Chapter 7 identify the opportunities to mitigate the barriers outlined in this chapter.

Proposal development, preparation, and submission are complex processes that take place within a broader sociocultural and sociohistorical context. Over time, these contextual factors have shaped the distribution of people and resources across the higher education and scientific industrial landscape on the basis of identity (e.g., gender, race/ethnicity, disability status). These factors also continue to structure opportunities available to historically underrepresented groups while also shaping the experiences of, perceptions of, and the interactions between various populations within science, technology, engineering, and mathematics (STEM). In this sense, many of the barriers that may limit the diversity of the pool of competed space mission PIs extend far beyond the proposal process itself. Thus, to identify opportunities to increase the diversity of the pool, it is critical for all relevant stakeholders—National Aeronautics and Space Administration (NASA), higher education institutions, the space science industry, and all members of the space science community—to understand and address these barriers.

[1] In June 2021, the National Academies contracted with NORC at the University of Chicago to design and complete a qualitative study of the experiences of scientists who have submitted space mission proposals to the NASA SMD. Data collection, which took place between August 2021 and September 2021, consisted of 29 individual, semi-structured interviews with scientists who had prepared and submitted at least one space mission to the NASA SMD as a PI from 2010-present. The final report of findings from the NORC study is included in Appendix C of this volume.

## ORGANIZING FRAMEWORK

A social ecological model consisting of nested, contextual layers is helpful to understanding the multiple factors that may impact demographic diversity among competed space mission PIs. The ecological model, originally developed by Urie Bronfenbrenner (1974, 1995) to understand human development, consists of concentric circles of influence and networks and has been applied to describe a broad range of social processes. In the context of this study, we adopt a social ecological framework to describe how factors that operate at the structural, institutional/organizational, interpersonal, and intrapersonal levels (see Figure 5.1) interact to yield the demographically homogenous pool of leaders of competed space missions that is currently observed. Although this model presents the levels as distinct, it is important to understand that this is a conceptual and analytic distinction. In practice, factors across these levels are intertwined and co-constitute one another (Risman 2004).

With this framework in mind, many critical questions emerge: Who is currently in the pool? How are the education-to-career pathways for distinct groups different? How are experiences within the field/institutions/process disparate? How do these differences limit access to resources? How does identity/background shape experiences within the field? How does identity/background shape access to collaborative networks/integration within science teams? Finally, how do one's own beliefs and biases affect decisions and behaviors related to team formation, proposal development, and proposal evaluation, and how does identity shape how one thinks about their work, likelihood of success, and tolerance for risk? Each of these questions relates to a potential barrier to diversity among competed space missions PIs. The following discussion of potential barriers draws on the social science and educational research literature on participation barriers for women and minorities in professional endeavors

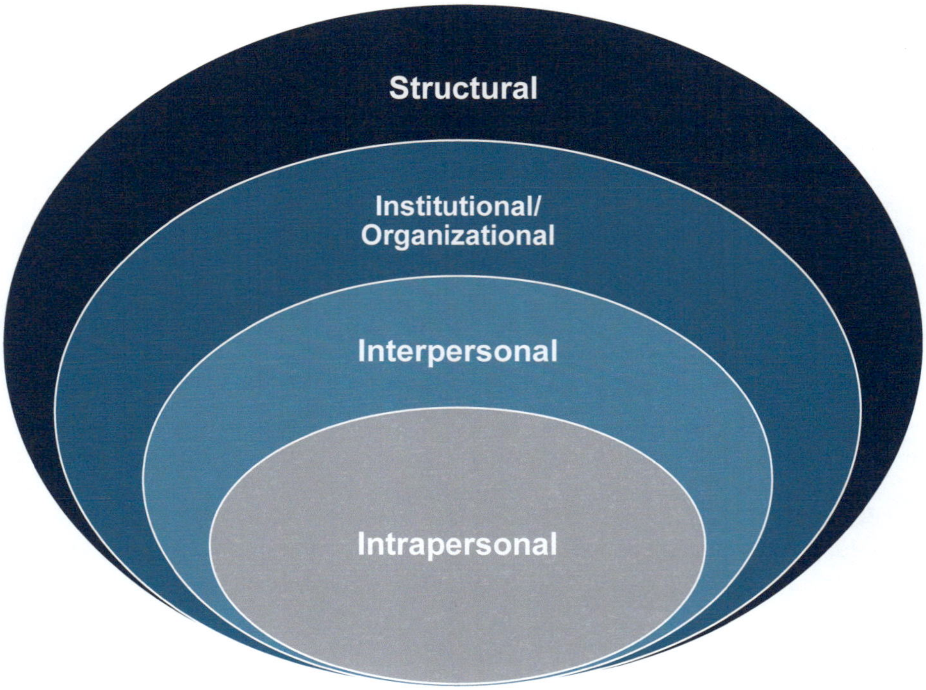


**FIGURE 5.1** Framework for understanding the interactional nature of factors impacting diversity and inclusion. SOURCES: See Bronfenbrenner (1995) and Sweat and Denison (1995).

such as science and engineering, as well as a commissioned qualitative study on the experiences of PIs with the current mission proposal process (see Appendix C).

In the social ecological framework depicted above, there are four nested and mutually reinforcing contextual layers (i.e., structural, institutional/organizational, interpersonal, and intrapersonal) that represent the environment in which a potential mission PI is embedded. The *structural* layer of the framework encompasses the systemic factors that originate from sociohistorical and sociocultural conditions and act to constrain access to power, resources, and opportunity on the basis of identity, resulting in advantages for some groups and disadvantages for others. These factors operate across all societal domains (e.g., education, economic, labor, health), and have implications for who has access to leadership opportunities on competed space missions. These encompass large-scale societal factors like public policy, ideology, lore, culture, history, systems of oppression, and deeply entrenched, widely shared gendered and racialized beliefs about which groups are capable of doing science and deserving of the opportunity to be educated and participate in STEM. Barriers that operate at the structural level shape the demographics of the pool of prospective PIs of competed space missions by constraining the pathways to and through STEM higher education and by shaping career opportunities (see Chapters 2 and 4).

The *institutional/organizational* layer of the framework includes the professional or academic environment in which a potential mission PI is embedded. Institutional and organizational factors comprise the individual workplace conditions which includes the resources and opportunities available to an individual in that workplace environment, implicit and explicit disciplinary norms, values and institutional culture, as well as institutional and agency practices and policies. These factors may have a disparate impact on prospective competed mission leaders from populations that have been historically excluded from and are underrepresented in the NASA SMD fields, and continue to be minoritized and experience marginalization (e.g., women, racially minoritized populations, people with disabilities, and members of the LGBTQ+ community). Institutional/organizational barriers often interact with structural, interpersonal, and intrapersonal factors to result in inequitable experiences for these individuals within their scientific field, the institutions in which they work, and the academic/institutional processes in which they engage. The differences in experiences within the field and within the discipline can also lead to inequities in the resources (e.g., money, staff, information, etc.) to which prospective PIs from historically excluded populations have access (see Chapter 2).

The *interpersonal* layer of the framework includes relational factors such as interactions (i.e., acts, words, and behaviors) between individuals and groups that can have a differential impact on historically underrepresented groups and erect barriers on the path to becoming a PI of a competed space mission. Identity—both social and personal—shapes relationships between individuals and among groups. Thus, prospective PIs with marginalized or minoritized identities will experience interpersonal interactions within the field differently. Additionally, identity can also shape access to collaborative networks and play a role in the degree to which an individual feels included or integrated within their discipline, department, or science team.

The final layer of the organizing framework comprises the *intrapersonal* factors that can act as a barrier to diversity in the leadership of competed space missions. Intrapersonal factors include the beliefs, biases, and thought processes that shape decision-making, the judgments we make about others, and how we view ourselves. For example, individual beliefs and biases can inform decisions and behaviors related to team formation, proposal development, and proposal evaluation, as well as those of women-identifying prospective PIs, racially minoritized groups or other historically underrepresented populations. Additionally, due to the interactions between the structural, organizational, and interpersonal factors mentioned above, identity can shape how individuals of different backgrounds think about their work, their likelihood of success, their tolerance for risk as well as how they prioritize specific experiences and goals over the course of their career.

## STRUCTURAL BARRIERS TO DIVERSITY

As with other STEM fields, women, racially minoritized populations, people with disabilities, members of the LGBTQ+ community and other groups that have faced systemic marginalization have been historically excluded from the educational and professional opportunities required to enter careers in the space sciences. Rooted in racism, sexism, and other systems of oppression, this intentional exclusion historically limited access to STEM learning, degrees, and careers to primarily White men. The vestiges of exclusionary policies and practices remain apparent today through the lack of diversity across all STEM disciplines, but particularly in physics, astronomy, and other related subfields (Ivie and White 2020; Mulvey 2021; NSF 2021d). Compounding past practices that limited access to space science careers for minoritized and marginalized groups, barriers along these educational and career pathways remain in place and continue to have a disproportionate impact on historically excluded populations (see Chapter 4). As a result, women of all races, racially minoritized populations, and other marginalized groups remain underrepresented among degree earners in the physical sciences (see Chapter 4) and in the space science workforce (see Chapter 3). This underrepresentation is even more acute among women of color, who experience the intersectional effects of racism and sexism (Cooper 2015; Cole 2021). Given this persistent underrepresentation in the physical sciences, the current pool of prospective PIs is largely homogenous. In this sense, a large structural barrier to increasing diversity among the leadership of competed space missions is the lack of diversity in the educational pathways to space science-related degrees and the resulting lack of diversity within the professional pathways to space science careers.

While the composition of the space science talent pool is a significant barrier to increasing diversity in the leadership of competed space missions, additional structural barriers limit access to opportunities to be involved in or lead competed missions for those minoritized and marginalized individuals who earn a PhD in the space sciences and take on faculty positions. According to the American Institute of Physics (AIP), women comprise a smaller proportion of faculty in PhD-granting physics departments than in departments where the highest degree granted is a bachelor's or master's degree (Helba et al. 2019). Because PhD-granting departments are typically present within well-resourced institutions with a high-level of research activity, the underrepresentation of women in these universities suggests that gender-based inequities in access to resource-rich institutional environments with PhD-level graduate students to support faculty research exist.

Previous surveys conducted by the AIP also found that Black and Latinx faculty in physics and astronomy are underrepresented in PhD-granting departments (Ivie et al. 2014). In 2012, the most recent year for which data are publicly available, half of Black physics faculty members worked at Historically Black Colleges and Universities (HBCUs), even though these institutions only comprise 4% of all physics departments in the United States (Ivie et al. 2014). With HBCUs less likely to offer graduate degrees in physics and astronomy than Historically White Institutions (HWIs), the concentration of Black physics faculty in HBCUs creates inequitable access to resources and graduate students to support their research. Indeed, 39% of Black physics faculty are employed at PhD-granting departments, compared to 49% of Latinx physics faculty and 60% of all physics faculty who work in PhD-granting departments (Ivie et al. 2014). Though they are more likely to be in a PhD-granting department than Black faculty, Latinx faculty remain underrepresented in PhD-granting physics departments compared to their White and Asian counterparts (Ivie et al. 2014).

The concentration of racially minoritized and women physics faculty in non-PhD-granting departments and the clustering of Black faculty in HBCUs is a structural barrier to advancing diversity among the leadership of competed space missions because these institutions are under-resourced and possess significantly smaller research infrastructures in the space sciences and other STEM fields compared to larger research universities. For example, in fiscal year (FY) 2020, approximately $5.33 billion in NSF funding and $1.74 billion in NASA funding were awarded to HWIs compared to the $0.0814 billion in NSF funding and $0.0153 billion in NASA funding awarded to HBCUs that same year.[2] These funding disparities need to be understood in the context of the fact that HBCUs granted nearly 22% of all bachelor's degrees in natural sciences and engineering earned by Black U.S. citizens and permanent residents in that same year.[3] These inter-institutional differences are longstanding and were purposefully engineered (Malcom-Piqueux 2020). The initial inequitable distribution of governmental investment across higher education institutions during the development of the nation's space science research and development (R&D)

[2] See NSF (2021b). Tabulations by committee using NCSES Interactive Table Builder available at https://ncsesdata.nsf.gov/home.

[3] Tabulations by committee using "IPEDS Completions Survey from Department of Education: 2019," in the NSF National Center for Science and Engineering Statistics Interactive Table Builder at https://ncsesdata.nsf.gov/home.

infrastructure created the current environment. Institutions who are more likely to have diverse STEM faculty (i.e., non-PhD-granting institutions, HBCUs and other Minority Serving Institutions [MSIs]) are also less likely to receive federal support for research programs that are related to NASA SMD fields (e.g., physics, astronomy). For example, in 2008, the most recent year for which these data are available, 85% of physics departments at HBCUs employed at least one Black and/or Latinx faculty member compared to 28% of physics departments at HWIs (see Figure 5.2), yet there are zero R1 HBCUs (Ivie et al. 2010).

The current disparities in R&D funding experienced by HBCUs and other MSIs originate in part from the way the federal government chose to invest in higher education institutions as it sought to expand the nation's research capacity during the Space Race. Between 1955 and 1965, federal R&D funding increased by 200% (Douglass 2000; Urban 2010), and a relatively small number of research universities received the lion's share of federal support for R&D in science, engineering, and technology. In 1963, twenty universities received 80% of all federal R&D funds awarded to higher education institutions, which enabled them to expand their research programs and strengthen their STEM research facilities, equipment, and other infrastructure (Douglass 2000). However, institutions that educated racially minoritized populations during this era did not enjoy the benefits of such investments that resulted in major improvements in research infrastructure elsewhere. Given that rate of growth in federal R&D investment during this period has not been matched in subsequent decades, the institutional inequities from this era are still apparent today. Currently, a larger number of higher education institutions receive R&D funding from federal agencies than did during the period of rapid expansion of federally financed research during the Space Race and the Cold War. However, a small number of institutions that benefited from the earliest investments continue to receive a significant proportion of federal R&D funds. In FY2020, for example, Johns Hopkins University (including its Applied Physics Laboratory) received 21.8% of NASA R&D funding awarded to higher education institutions, while the nation's HBCUs received a combined share of just 0.87% of NASA R&D funding granted to higher education institutions (NSF 2021b). Neither past nor current R&D funding levels reflect the disproportionately large role that these institutions play in educating Black scientists (NSF 2021d) and employing faculty of color (Ivie et al. 2010; McCrary 2021). This historical underinvestment in the space science R&D infrastructure at HBCUs and other MSIs has in effect limited the capacity at these institutions to compete for a NASA space mission. Moreover, given the racialized patterns of where today's faculty are employed and where degree earners in Science Mission Directorate (SMD)-related fields are educated, historical underinvestment

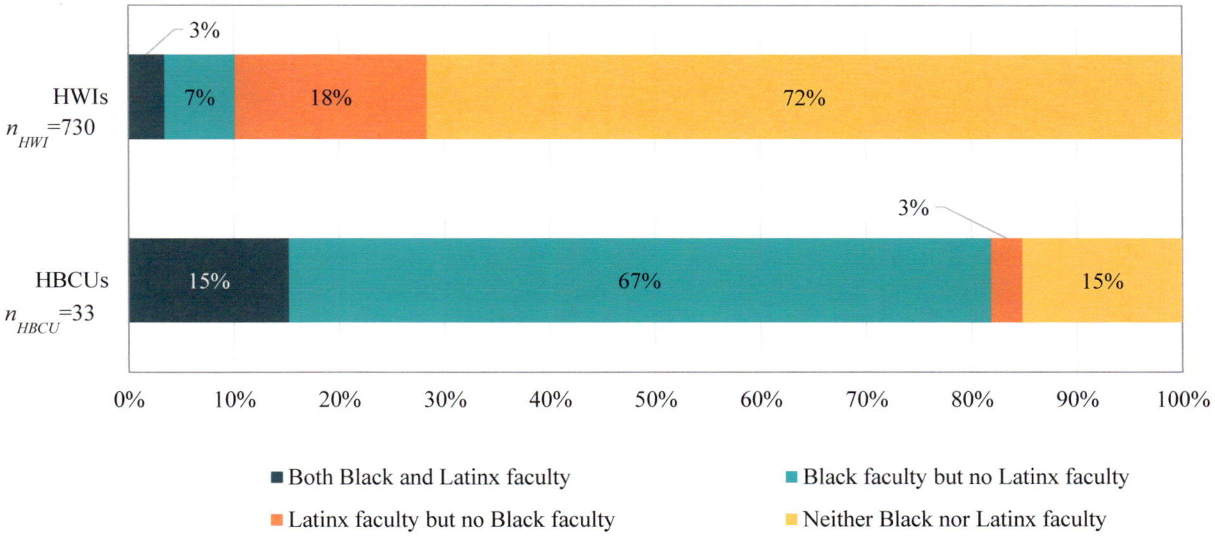


**FIGURE 5.2** Proportion of physics departments at HBCUs and HWIs with African Americans and Latinx faculty in 2008. SOURCE: Data from Ivie (2010).

leads to a disproportionately negative impact on faculty members of color in SMD-related fields and limits access to early-career mission experiences for students of color at MSIs.

Findings from the commissioned qualitative study of scientists who submitted competed mission proposals to NASA underscore the value of early career mission experience, which was often gained through connections to faculty mentors with NASA funding during undergraduate and graduate school training. MSIs and non-PhD-granting institutions that receive less NASA funding than HWIs (NSF 2021c) are at the same time also educating a disproportionately large share of racially minoritized students in physics, astronomy, and other SMD-related fields. These funding disparities likely reduce access to the type of early career mission experiences that PIs identified as critical to their training and preparation to become a PI.

The commissioned qualitative study also revealed the importance of the institutional research infrastructure to a successful competed mission proposal submission. Respondents cited the vitally important role of the site visit during the proposal submission process and expressed the significant amount of preparation and institutional resources and support necessary to plan and execute a successful site visit. Given that women physics faculty and Black and Latinx physics faculty are more likely to be employed at non-PhD-granting institutions, HBCUs, and other MSIs compared to White and Asian male physics faculty, the modest research infrastructures, lower levels of research funding and administrative structures to support STEM research at these institutions,[4] are barriers that disproportionately impact faculty from historically underrepresented groups as they pursue leadership positions on competed space missions.

**Finding:** Non-PhD-granting academic institutions, HBCUs, and other MSIs play a disproportionate role in educating physics degree earners and employing physics and astronomy faculty who are women and/or from racially minoritized groups. Due to long-standing disparities in research capacity and infrastructure, these institutions receive less NASA funding than their R1 peers.

**Conclusion 5-1:** Non-PhD granting academic institutions, HBCUs, and other MSIs experience inequitable access to the mission experiences that are helpful to scientists who go on to submit competed space mission proposals.

**Conclusion 5-2:** Proposal development, preparation, and submission—including site visits—are resource-intensive processes. Given the concentration of women physics faculty and Black and Latinx physics faculty in less-resourced institutions, resource-intensive processes likely disadvantage prospective PIs from these underrepresented populations.

## INSTITUTIONAL AND ORGANIZATIONAL FACTORS

Disciplinary culture and climate within Earth sciences, planetary sciences, astrophysics, and heliophysics and at PI institutions (academic and non-academic) can act as barriers to greater diversity among the leadership of competed space missions. In this context, culture refers to the implicit and explicit customs, behaviors, norms, and values that are deeply embedded within the discipline or field. The values and norms of STEM disciplinary culture have been explored as contributors to the unwelcoming environments often experienced by women and minoritized populations in these fields (Griffin 2018; Posselt 2020). Climate, which encompasses psychological and behavioral dimensions (Hurtado et al. 1998), refers to how individuals and groups experience their field, institution, or department as well as the quality and extent of interaction between those groups and individuals (Rankin and Reason 2005). Within STEM fields, an exclusionary climate and perceptions that one's own value and identity are incongruent with STEM disciplinary culture are often experienced by historically excluded, underrepresented, and minoritized groups (Griffin 2018). Culture and climate can affect prospective PI engagement, sense of belonging, persistence, and performance

[4] See the Finance Survey Component of NSF (2021c); data available at https://nces.ed.gov/ipeds/use-the-data.

Tenure policies that incentivize a high volume of publications can also act as an institutional/organizational barrier to diversifying the pool of PIs. Given the demographic makeup of the SMD fields (e.g., a higher proportion of women in the Division for Planetary Sciences (DPS) of the American Astronomical Society (AAS) survey were in pre-tenure, tenure-track positions than in tenured positions compared to the proportion of men) and the structure and timeline of the mission proposal development process (typically 2-3 years in length), prospective PIs may be required to de-prioritize other shorter-term projects that would lead to publications. Thus, there is great risk involved in this endeavor because of the significant chance that the mission will not be selected (see Chapter 2), and this risk may be even greater for aspiring PIs at the early career stage where publications are important productivity indicators for reappointment. This disproportionately disadvantages those from minoritized communities.

The commissioned qualitative study also points to potential disparities in how proposals may be reviewed given the lack of a double-blind process. At the proposal stage, there are many factors to consider and the PI's status and reputation were among those cited by respondents as having implicit prestige bias because of the way in which the reputation of the PI may implicitly influence the review panel. The respondents believe the organizational culture of NASA emphasizes the need for a PI to be well respected by scientists in the field and industry partners, while also being "taken seriously" by those outside of the agency that may interact with the PI (e.g., policy makers, media). Therefore, previous success as a PI or previous collaborations with well-respected scientists may be an invisible criterion not formally adopted as part of the internal review process. This idea that only a select group of individuals are qualified to become PIs may reinforce the status quo maintaining the same group of PIs to dominate the competed space missions. Respondents also believed that the feedback they received on why their proposal was not selected did not seem to reflect some of the perceived hidden criteria regarding PI status and reputation.

The literature points to the complicated nature of determining whether bias is present in proposal review processes (Sato et al. 2021). The effect of such biases on funding decisions are difficult to investigate due to the opacity of many federal agencies' review processes, including the demographic composition of review panels and access to review panel discussions and deliberations. While there is evidence of racial disparities in the National Institutes of Health (NIH) R01 research awards after controlling for educational background, prior success, and employer characteristics (Ginther et al. 2011) and researchers have identified the decision points in the NIH review process at which biased assessments emerge (Erosheva et al. 2020), it is not understood how these potential biases operate to shape the assessments and judgments of reviewers and review panels resulting in the observed disparate funding outcomes. Previous research highlights the multiple factors including individual and systemic biases, social barriers, and methodological challenges associated with collecting and analyzing bias free data. Additionally, the decision-making process for proposal selection is often ambiguous and difficult to account for all elements considered by reviewers. Though more research is needed to understand how bias manifests during the review process, a widely shared perspective among respondents in the commissioned qualitative study was that bias is a barrier in the current process.

Another institutional and organizational factor is the perceived advantage that NASA center researchers have over those not affiliated with a NASA center. Respondents described the ways in which NASA center scientists may have a better understanding and alignment with the NASA reporting requirements, budget monitoring, and administrative processes. The respondents identified a need for administrative support for proposers' institutions and noted that there is variability in who is receiving this assistance.

## INTERPERSONAL FACTORS

An important mechanism producing differential and long-lasting effects on the careers of potential and current PIs is their treatment as mentees. Respondents in the commissioned qualitative study and PIs describing their experiences during open, information-gathering sessions at committee meetings underscored the importance of having supportive mentors for their ability to access critical experiences with missions. PIs who described having mentors who were supportive and encouraging described enriching training that exposed them to elements of space missions and allowed them to gradually increase the complexity of their responsibilities.

Not all PIs had access to these positive mentor-mentee relationships, and among the respondents, access followed gendered and racial lines. For example, women and racially minoritized respondents were more likely than White men respondents to report that they received very little support from mentors, and even overt hostility. Negative experiences with mentors included being passed over for opportunities for which they were qualified and feeling "stuck" in toxic mentor-mentee relationships that then affected the climate of the work environments.

This lack of positive mentoring experiences can have cumulative effects: PIs who entered the space mission community after graduate school reported feeling perpetually "behind" peers in terms of grants and other professional accomplishments. Lack of access to collaborative networks can also act as a barrier to diversifying the pool of PIs of competed space missions for women and scientists of color. For example, in a study of graduate students in physics and astronomy, women, on average, rated their doctoral advisors less positively than did men (i.e., were less likely to agree that their advisor encouraged them in their research and career goals, provided adequate input, or was easy to discuss ideas with) (Porter and Ivie 2019). Women graduate students in physics and astronomy were also more likely to reporting having a mentor outside of their advisor during graduate school (Porter 2019). Among women in astronomy, those who reported a lower quality of relationship with their doctoral advisors were more likely to leave the field after graduating (Porter 2019). There is also evidence of lower access to mentors for women and scientists of color after PhD completion and entering a space science career. In a 2020 AAS DPS Climate Study, women were more likely to cite lack of access to mentoring from senior colleagues as a negative influence on their career satisfaction (Porter et al. 2020). In the same study, Black, Latinx, and Asian scientists were more likely to cite a lack of interactions with well-established scientists as an impediment to their careers (Porter et al. 2020). These studies of the space sciences community did not explore differences in the quality of mentoring relationships based on mentor-mentee "identity match" (e.g., race, gender). However, research from other disciplines suggests that while homophily is no guarantee of a successful mentoring relationship, sharing a salient social identity can support the connection that a protégé feels with their mentor (Baker et al. 2014). Given the value of multiple mentors to mentees' career advancement and growth (NASEM 2019b), limited access to mentoring is a barrier that disproportionately impacts populations that are underrepresented within the space sciences. Considering the current demographic reality within STEM fields broadly and within the space science community more specifically, encouraging the adoption of inclusive mentoring practices by all mentors—regardless of identity—may mitigate the barriers experienced by scientists from underrepresented populations. Examples of inclusive mentoring practices include: listening actively, moving beyond "colorblindness," intentionally considering how culture-based dynamics like imposter phenomenon can negatively influence mentoring relationships, and reflecting on how biases and prejudices may affect mentoring relationships (NASEM 2019b).

In the context of competed missions, respondents in the commissioned qualitative study revealed the impact that differential access to networks of PIs could have on the potential PIs' ability to access the necessary resources, personnel, and informal knowledge to prepare a competitive proposal. From the perspective of researchers who aspired to be PIs, but were not integrated through their graduate training or through the sponsorship of a collaborator, the scientific community engaged in competed space missions were viewed as nearly impenetrable. Aspirant PIs not incorporated into these social networks had to go to great lengths to gain entry including cold calling, scoping out "power brokers" at conferences, and otherwise investing a great deal of time and effort just to have access to the social network.

Prior research on network access in STEM has found that women, people of color, and LGBTQ professionals face greater challenges of being seen as credible or garnering respect within scientific networks (Moss-Racusin et al. 2012; Eaton et al. 2019; Cech and Waidzunas 2021). This greater challenge establishing credibility, alongside less access to positive and supportive mentorship, means that marginalized and minoritized population are at a significant disadvantage gaining access to the social networks critical to developing the skills and experience necessary to prepare and lead competed missions. This differential network access can aggravate another form of differential treatment identified in the commissioned qualitative study: a perceived bias by NASA toward experienced PIs over new PIs. While NASA has indeed used previous PI experience as an explicit criterion in the past,[5] the respondents perceived that this process has continued. This has implications for new PIs generally, but is particularly pertinent for the lack of diversity among successful PIs. Because younger cohorts of scientists are

[5] See, for example, the SMEX 2007 AO available at https://explorers.larc.nasa.gov/CommAnnouncementFAQ.html.

more demographically diverse than older generations (NSF 2021d), any bias toward more advanced-career PIs would produce a less diverse set of awardees than the pool of applicants.

Research evidence has also shown that homophily[6] is common in research collaborations, with academic institutions attended and genealogy (individuals with whom one trained/studied/worked) shaping scientists' primary collaborative units (Freeman and Huang 2014). Additionally, teams are often constructed based on areas of expertise, leading to the construction of professional connections and networks based on shared scientific interests. Given the racialized and gendered nature of academic and career pathways, gaining access to professional networks may be a larger barrier to gaining mission experience for women and scientists of color. This is particularly true if prospective PIs rely on their existing networks when forming collaborations for subsequent funded projects instead of looking for new team members (see Chapter 2).

To gain access to collaborative networks, well-established scientists (who are disproportionately White and male) must make judgments, assessments, and decisions about early career scientists' abilities, qualifications, competence, and suitability to join their team. These assessments are subject to bias, defined as attitudes, behaviors, and actions that are prejudiced in factor of or against one person or group compared to another. Biases may be held by an individual, group or institution, and can lead to unfair advantages or disadvantages. Biases can be explicit (i.e., conscious), or implicit (i.e., unconscious). Implicit bias occurs automatically and unintentionally, but affects an individual or group's judgments, behaviors, and decisions (Olson and Fazio 2003). A great deal of evidence illustrates how, across a range of contexts, biases can lead to inequitable assessments based on widely shared stereotypes and implicit associations that are highly racialized, gendered, ableist, etc. (Goldin and Rouse 2000; Fiske et al. 2002; Bertrand and Mullainathan 2004; Eaton et al. 2019). Given deeply entrenched social hierarchies and systems of power and privilege, implicit biases can have a disproportionately negative effect on women, racially minoritized individuals and other marginalized groups, the very populations that have been historically excluded from and remain underrepresented in STEM fields, including the space sciences.

The commissioned qualitative study also revealed patterns of biased treatment in teams within the context of competed space mission teams. As researchers were amassing experience on mission teams as Co-Investigators, some reported biased treatment and hostile environments by respective PIs. Respondents reported experiences where PIs were domineering, tended to micromanage, and fostered "old boys clubs" that implicitly or explicitly excluded women and other minoritized groups. Even once they held PI positions, women and racially/ethnically minoritized scientists shared a range of experiences such as being the targets of sexist jokes, being labeled as "difficult" to work with even when performing standard PI job functions, and having their basic competence questioned. Additionally, respondents reported that when women and racial/ethnic minority PIs sought to establish their leadership, they were more often seen as "arrogant," while White men with the same characteristics would be viewed as "confident."

**Finding:** Prospective PIs perceive multiple forms of bias throughout the proposal development, preparation, and submission processes, and report having experienced discrimination even after assuming the leadership position for a successful competed mission.

**Finding:** On average, members of the space science workforce from historically excluded and underrepresented groups (i.e., women, scientists of color) neither enjoy the same access to collaborative networks nor the benefits of mentoring relationships as individuals without marginalized identities.

**Conclusion 5-3:** Mentoring and access to collaborative networks that include experienced PIs are formative in preparing those who submit competed space mission proposals. However, women and racially minoritized space scientists report less access to mentors and a lower quality of relationship with their doctoral advisors

[6] Homophily refers to people seeking out persons similar to themselves. For example, a researcher may seek out another researcher with similar social identities to themselves like a White cis-gendered male researcher seeking another White cis-gendered male researcher to be part of their research team or as a suitable mentee.

and senior colleagues. Inequitable access to high-quality mentoring relationships is a barrier to increasing diversity in the leadership of competed space missions.

## INTRAPERSONAL FACTORS

Intrapersonal factors can act as barriers with a disparate impact on scientists from historically excluded and underrepresented groups. For example, developing a sense of belonging (i.e., do I feel that I am a fully accepted and engaged member of my scientific community) may be more difficult for individuals from minoritized and marginalized groups, especially if they are one of a few in that environment (Rainey et al. 2018; Fisher et al. 2019; Xu and Lastrapes 2021). Prior research has robustly connected a sense of belonging with a host of outcomes, including intentions to stay in one's scientific field long term (Lewis et al. 2017).

Development of a sense of belonging is hindered by the unequal access to supportive and opportunity-rich mentoring that was described above. In the commissioned qualitative study, respondents who experienced positive mentorship experiences reported that these were foundational to their early sense of belonging in the field. Effective mentors introduced their mentees to "powerful players" in their network and helped them decipher the mission proposal process and understand proposal timelines. This early exposure to missions gave early-career scientists time to develop confidence and a sense of identity with space missions that bolstered their ability to endure the uncertain and challenging process of submitting a proposal as PI.

Additionally, imposter phenomenon, or the psychological experience of professional or intellectual fraudulence (Clance 1985; Sakulku 2011; Chakraverty 2019), due to the widely shared assumptions about who is capable of being a PI. The language used to communicate the expectations of what it takes to be a PI (e.g., being hypercompetitive) may fuel feelings of intellectual and professional fraudulence, as women and scientists of color are acutely aware of the deeply rooted stereotypes about their respective groups. Even when equally qualified and experienced, lack of confidence can lead minoritized researchers to avoid applying for competitive funding and awards and even to consider leaving their STEM field entirely (Smith et al. 2013; Tao and Gloria 2018; Abrica et al. 2020). Imposter phenomenon is closely tied to a broader concept of "self-efficacy," or one's judgment about their ability to execute the necessary tasks to attain a goal. Due to structural and interpersonal processes like those described above, past research has shown that minoritized scientists often develop lower science-related self-efficacy, which can, in turn, be tied to differentials in persistence and confidence (Larose et al. 2006; AAUW 2008; Rittmayer and Beier 2008; Majer 2009).

Study respondents indicated several specific parts of the proposal process that undermined their confidence. For instance, minoritized researchers were more likely than White men to encounter challenges to their competence as they attempted to access networks. Even when serving as PI, women and people of color were more likely to report having their expertise questioned by their team members. This undermining of confidence is something that White men PIs are less likely to have to deal with and is an important mechanism in the decision to seek out mission leadership at all. For some respondents, their experiences in science mission teams produced the strongest and clearest experiences of discrimination or bias of their entire careers.

These direct and indirect refutations of minoritized scientists' competence is aggravated by general cultural lore among the space science mission community that very few have what it takes to serve as PI (Shapin 2008; Sheltzer et al. 2014; Leslie et al. 2015; Posselt 2020). This is a gatekeeping belief that not only may prevent community members from recognizing mission leadership potential in women and people of color, but also hinders women and scientists of color from seeing themselves as capable of such leadership.

Among those women and people of color who do pursue mission leadership, experiences with tokenism may also have play a chilling effect on their desire to take on the PI role in future proposals. Tokenism is the act of making only symbolic efforts to increase representation of persons from minoritized groups, particularly by recruiting one or two people from those groups to give the appearance of equality within an organization or work group (Kanter 1977a; Roth 2004; Holgersson and Romani 2020). Tokenism is especially common in spaces where women and people of color have faced particularly entrenched historical exclusion (McIlwee and Robinson 1992; Blair-Loy and Cech 2022).

The commissioned qualitative study revealed several instances where minoritized scientists believed they were only elevated to PI role because of the "diversity" they brought to the team, not principally because their team members respected their leadership. For instance, one PI reported that they were "used" by the mission team as a token to make the other members of the mission team appear to care about diversification of leadership, but later learned that these other team members knew even before submission that their proposal was considered a long-shot and was not ultimately funded.

A final intrapersonal barrier is the tradeoffs that potential PIs may see in proposal preparation and submission. The overwhelming consensus of career narratives described in presentations to the committee and the results from the commissioned qualitative study is that proposal preparation is intensely demanding and difficult, if not impossible, to combine with non-work responsibilities like childcare. The majority reported that proposal preparation was "extremely taxing" on their personal lives. Many explicitly noted that they were only able to devote the time necessary for proposal preparation because they did not have children or they either had supportive partners who took on the additional burdens of childcare and housework or they were able to hire out many tasks (e.g., personal shopper). Because women professionals are more likely than their male peers to be shouldered with primary responsibilities for childcare in their families (Ecklund et al. 2012; Cech and Blair-Loy 2019), and professionals of color are more likely than White professionals to have eldercare and other family care responsibilities (Pinquart and Sörensen 2005; FCA 2012), they are less likely to have the privilege of downshifting these personal responsibilities throughout the award process. In this sense, the proposal process, as currently designed, can be a "heavier lift" for prospective PIs with minoritized identities.

**Finding:** Prospective PIs from historically excluded populations reported that interpersonal interactions with others in their field undermined their sense of belonging and fueled experiences of imposter phenomenon, making the path to leadership of competed space missions more difficult. Similarly, incongruence between the messages about "what it takes" to be a PI and prospective PI's self-perceptions erected barriers along the pathway to competed space mission leadership positions.

**Conclusion 5-4:** Experiences of bias, discrimination, and exclusionary interpersonal interactions hinder the creation of an inclusive working environment and act as a barrier to diversity among the leadership of competed space missions.

The chapter attended to barriers within different levels of an environment; the factors at each level should be understood as often mutually reinforcing and co-constitutive. Factors at one level may presuppose and help reinforce factors at another. It is important to note that the barriers identified and discussed in this chapter are not exhaustive, and further study is needed to substantiate these preliminary findings and understand the full suite of barriers that influence the resulting makeup of the prospective leadership pool for competed missions. Such steps are crucial for identifying those points in the proposal development, preparation, and submission process (see Chapter 2), the work environment, and educational and career pathways (see Chapter 4), where intervention may be needed. Best practices and additional recommendations for mitigating these barriers are offered in Chapters 6 and Chapter 7.

# 6

# Examining Best Practices at the Federal Level for Advancing Diversity, Equity, Inclusion, and Accessibility in Principal Investigator–Led Missions

Throughout this report the committee has attempted to delineate the end-to-end mission proposal process; the extent of demographic diversity within the National Aeronautics and Space Administration (NASA) Science Mission Directorate (SMD)-related fields, workforces and proposer pools; as well as the variety of factors underlying these realities. In the course of examining these areas, several issues emerged that warrant further attention in an effort to articulate clear paths forward for NASA SMD, as it relates to realizing its stated goal to increase diversity, equity, inclusion, and accessibility (DEIA) in the leadership of competed space missions. A growing body of literature, as well as an increasing number of examples from other institutions and organizations have put forth strategies and practices that NASA SMD can either adopt or adapt to make progress in a number of areas that can aggregately lead to the desired outcomes for principal investigator (PI)-led missions.

In the charge to the committee, other federal agencies, such as the National Science Foundation (NSF) and the National Institutes of Health (NIH), were specifically cited as references to consider in providing guidance for overcoming barriers to diversity and inclusion in the context of funding large, PI-led, multi-institutional projects. In looking across the landscape of federal agencies, the committee found that other organizations, such as the Department of Energy (DOE) and the Department of Defense (DoD), utilize vehicles similar to the NASA announcement of opportunity (AO) when procuring large-scale goods and services, albeit not particularly well defined or understood as can be the case with new technologies. In this chapter, we examine the policies, processes, and education/development programs at representative federal science agencies and organizations. The committee also examines the extant literature, and in the committee's judgment, present the practices that are currently considered the best to foster DEIA in the following areas: (1) proposal review, (2) data collection, (3) accountability, and (4) investment in science, technology, engineering, and mathematics (STEM) pathways.

## DEFINITIONS OF BEST PRACTICES

The term "best practice" is widely used in a number of sectors such as business, industry, education, research, and public policy; but often used without an accompanying definition for what this designation actually means in the respective context. In a recent report, *Best Practices for Diversity and Inclusion in STEM Education and Research: A Guide for and by Federal Agencies*, a "best practice" is defined as "a procedure that has been shown by research and experience to produce optimal results and that is established or proposed as a standard suitable for widespread adoption" (NSTC 2021). Additionally, the report outlines three levels to distinguish among best practices based on the extent of the evidence-base supporting the effectiveness of the approach: *evidence-based practices, promising practices, and emerging practices*; as well as definitions of each term in order to standardize their use and improve overall reporting of best practices by federal agencies.

*Evidence-based practices* are defined as those that are grounded in research and evaluation and have met some established test of validity in improving a specific outcome. *Promising practices* are those that have been successfully implemented, but sufficient evidence that substantiates all the parameters associated with success of the practice has not been collected or generated. Additionally, practices that are known to be "evidence-based" in one context, would be defined as "promising" when being applied in a different context. Finally, *emerging practices* are those that are new, innovative, or exploratory in nature, and while they may be based on some level of evidence, it is not sufficient for the practices to be considered "promising" or "evidence-based." In making determinations about relevant policies, processes, and programs at federal agencies for addressing DEIA, and for consistency with designations already in place across federal agencies, the committee adopted these definitions in the context of this report. Additionally, it is important to note that the committee did not identify examples of evidence-based, promising, and emerging practices in every area discussed.

## FOSTERING DEIA IN PROPOSAL REVIEW FOR SCIENCE AND ENGINEERING RESEARCH ACTIVITIES

In Chapter 2 of this report, we described the comprehensive process through which proposal review for competed space missions occurs as governed by section 1872.4 of the NASA Federal Acquisition Regulation (FAR) Supplement. The needs in terms of what is required to put together a winning proposal (i.e., compelling science investigation, technically feasible, and programmatically credible) vary across mission scope and size, and becoming a PI or co-investigator (Co-I) on a winning proposal is typically a long training and development process for the investigators as well as the proposing team as a whole. The resources required for such proposal development in terms of dollars and time are not negligible. Neither are investments for the awarded projects negligible, ranging from tens of millions of dollars—over $1 billion awarded funds to winning proposals. Thus, the need for such extensive resources creates a challenging environment for both proposers and reviewers. Yet, despite efforts on the part of NASA to ensure that this stage of the proposal process is carried out responsibly and with fairness and objectivity, the influence that implicit and explicit bias can have on the results of proposal review are significant (see Chapters 2 and 5).

To this end, NASA is not the only agency facing this challenge, and in fact both NSF and NIH have piloted experiments to identify and reduce the impact of any possible bias in grant proposal evaluation in response to findings of an uneven playing field between proposers (EOP and OPM 2016). In the case of NIH, substantial funding disparities were found to exist between African American/Black and White applicants, even when controlling for demographics, education, training, employer characteristics, NIH experience, and research productivity (Ginther et al. 2011). African American/Black scientists were significantly less likely to receive Research Project Grant (R01) funding compared to White scientists. Some important factors that were found to be associated with the observed funding gap included assignment of poorer scores and decisions to discuss an application or not—factors that are at the hands of reviewers (Hoppe et al. 2019; Erosheva et al. 2020). Clearly R01 grants and large acquisitions such as NASA space missions selected through a competitive process that is governed by FAR are not equivalent, and so this case is only meant to be an illustration of the impact that possible bias can have in the context of federal research funding, and possible measures to take in assessing and addressing it.

### Promising Practices for Assessing the Proposal Review Process

To understand whether bias in proposal review contributes to the observed funding disparities, the NIH has employed multiple promising approaches. These include testing the utility of anonymizing grant applications before review; conducting more research to gain a richer understanding of scientific, technical, and demographic issues that might be at play; using text-mining software to examine applications and reviewer critiques for evidence of potential bias; reexamining the evaluation process used to review grants for indicators of bias on a continual basis;

and providing opportunities for early-stage researchers to serve on review panels (Valantine and Collins 2015; Nakamura et al. 2021). Additionally, the NIH Office of Extramural Research is engaging in several analyses of applicant flow data to examine differences between groups (e.g., according to race/ethnicity, disability status, etc.) in their matriculation through the proposal process from application to selections.

NSF's Merit Review process is a mechanism by which the agency ensures that the competitive peer review of proposals submitted for support identifies proposals that are of high-quality and advance scientific knowledge, and is carried out with biases and conflicts minimized. Additionally, this process ensures that proposers are provided constructive and useful feedback. As part of proposal review, reviewers receive a short orientation that covers topics such as writing analytic reviews, the broader impacts criterion, and training on ways to reduce the impact of unconscious bias including anchoring bias, confirmation bias, halo effect, and language bias (NSF 2020). Additionally, an external Committee of Visitors periodically examines the integrity and efficiency of the merit review process, and the resulting reports along with NSF responses are made publicly available. With these strategies in place, a 2019 Survey of proposers and reviewers indicated that 74% of survey respondents agreed their proposal was treated fairly in the review process, despite the fact that 73% of their proposals could not be funded. Moreover, the impact of these strategies on the participation of different groups in the agency's award portfolio is documented and openly reported on an annual basis.

### Promising Practices for Proposal Evaluation

A substantial amount of research evidence demonstrates that diversity and inclusion positively impact innovation and the quality and outputs of science (Cohen et al. 2002; Page 2007; Bell et al. 2011; Hofstra et al. 2020). In fact, much of the legislation around broadening participation in STEM during the Obama administration was motivated by the acknowledgment that enriching the STEM enterprise and the innovations and solutions it produces, is inextricably linked to realizing the potential contributions of people from all backgrounds (EOP 2016).

In 1997, NSF established a specific broader impacts criterion as part proposal review (NSF 1997), which evaluates how the proposed work will advance scientific knowledge, and also contribute to the achievement of societally relevant outcomes. Such outcomes include, but are not limited to full participation of women, persons with disabilities, and underrepresented minorities in STEM; improved STEM education and educator development at any level; increased public scientific literacy and public engagement with science and technology; improved well-being of individuals in society; development of a diverse, globally competitive STEM workforce; increased partnerships between academia, industry, and others; improved national security; increased economic competitiveness of the United States; use of science and technology to inform public policy; and enhanced infrastructure for research and education (NSF 2008). The committee could not find any documented evidence of direct impact as a result of requiring this criterion, because NSF's merit review process considers intellectual merit and broader impacts together when evaluating the impact of its federal investments. However, some indirect impacts have been captured through the outcomes of funded projects: economic impact via the development of new billion-dollar industries[1] and societal impact via the establishment of a community of practice for developing sustainable and scalable institutional capacity and engagement in broader impacts activities.[2]

Other agencies have also implemented diversity and inclusion criteria as part of proposal evaluation. In 2017, the National Institute of General Medical Sciences (NIGMS) made diversity at all education and career levels a part of the scored review criteria (Gammie 2021). In funding its Cooperative Institutes (CIs), the National Oceanic and Atmospheric Administration (NOAA) has also established evaluation criteria that assesses proposal plans to recruit and train a more diverse and inclusive workforce, as well as to involve partnerships with other universities or research institutions, including Minority Serving Institutions (MSIs) and NOAA Educational Partnership

[1] Efforts through the Computer Science and Telecommunications Board at the National Academies, which is funded by NSF, at https://www.nationalacademies.org/our-work/depicting-innovation-in-information-technology.

[2] Efforts through the National Alliance for Broader Impacts (NABI) Center for Advancing Research Impact in Society (ARIS), at https://researchinsociety.org.

Program (EPP)/MSI Cooperative Science Centers, in ways that describe how the engaged partners will meaningfully contribute to the proposed activities of the CI (Koch 2021).

The committee acknowledges that NASA provides unique opportunities to PIs to run incredibly large budget projects (tens of millions of dollars to over $1 billion), and recognizes that no other federal agency offers comparable science and engineering (S&E) investments in their acquisitions for similar purposes. Nevertheless, NSF's Materials Research Science and Engineering Centers (MRSECs) were identified in the study's statement of task as the closest in nature to NASA's PI-led missions in terms of the proposal review process. It should be noted that other NSF programs like the Physics Frontier Center and the Science and Technology Centers programs follow a similar format and at higher budget levels. Looking specifically at MRSECs for illustrative purposes, they comprise two to three topically distinct core research units that operate as interdisciplinary research groups (IRGs), each IRG employs interdisciplinary research to address unique, fundamental materials research problems that are intellectually challenging and important to society, and each are supported through 6-year awards ranging from ~$1.6 million to ~$4 million per year (Jones 2021). Proposal review for MRSECs is comprised of three stages: (1) pre-proposal (similar to step 1 of two-step mission proposals), (2) full proposal (similar to step 2 of mission proposals during which down selection occurs), and (3) site visits (similar to the final stage of selection for some mission proposals). Additionally, during the proposal review process MRSECs are partially evaluated based on their Diversity Strategic Plan, which must (1) articulate plans to broaden participation at all levels; (2) demonstrate a meaningful inclusion of participants from underrepresented groups: faculty, post-docs, graduate and undergraduate students, and staff; and (3) include metrics for assessment and desired, attainable outcomes. Moreover, implementation and efficacy of the Diversity Strategic Plan are addressed during site visits and at the time of annual reviews.

## Emerging Practices for Proposal Evaluation

At the time that the work of this committee began, considering diversity on mission proposal teams was encouraged, but no specific evaluation criteria for diversity existed in the context of proposal review. In September 2021, NASA released a request for information (RFI) on "Adding Inclusion, Diversity, Equity, and Accessibility Requirements to NASA Announcements of Opportunity" (NASA 2021h). This RFI announced NASA's desire to add explicit language about DEIA to the standard AO template, all future SMD AOs and the Stand-Alone Missions of Opportunity (SALMON) AO (see Annex 6.A for more details on this AO type), and amend a SALMON AO that was open at the time of the release of the RFI (NASA 2021g). As part of this new policy, proposals are required to

- "Include a description of the processes used to assemble the proposed team and how these processes align with NASA's IDEA values," and
- "Describe the processes that will be employed to enable and monitor (i) the creation and maintenance of an inclusive and equitable environment throughout the project lifecycle, (ii) the maintenance of a diverse team, (iii) and the continued access to equitable opportunities for contributions from team members toward mission success."

More importantly, these DEIA plans would now be assessed as part of the evaluation of Scientific Implementation Merit and Feasibility of the Proposed Investigation (also known as Form B; see Chapter 2 and Annex 2.B, Table 2.B.2); however, no separate score or grade would be provided for this component.

It remains to be seen what outcomes such policies will have on the pool of PIs for competed space missions, but the committee acknowledges that this is an important step taken by NASA and commends the agency for this effort. Furthermore, if NASA is able to pivot successfully with the implementation of the new DEIA requirements for mission proposals, then the agency is positioned to be an exemplar for other organizations who utilize similar procurement processes, but have yet to adopt such policies. We provide additional considerations for utilizing diversity criterion in Chapter 7.

**Finding:** Recently, NASA has taken steps to require and evaluate DEIA plans as part of its AOs for competed missions, but compared to NIH and NSF, fewer approaches have been employed by the agency to identify and reduce any possible bias in the current mission proposal process.

## DATA COLLECTION AND PROMOTING DEIA IN S&E FIELDS

Federal agencies like NASA play a significant role in supporting S&E in the United States, and in 2020, total federal research and development (R&D) funding was estimated to be to $155.6 billion (NSF 2021c). Therefore, such agencies have an obligation to be wise stewards of investing taxpayers' dollars, and to be effectively transparent about any knowledge ascertained in S&E fields whenever possible, thus providing evidence of an appropriate return on the public's investment. As mentioned in the previous section, some federal agencies have established systematic processes for tracking and assessing impacts on investments. In the case of NASA, there is evidence that within the Office of STEM Engagement (OSTEM) a comprehensive performance and assessment strategy is in place for monitoring and reporting accomplishments, particularly progress toward pre-established goals, and for collecting and analyzing data to assess how well a particular initiative is working and why. But as mentioned in Chapter 3 of this report, there is currently no adequate mechanism in place for collecting accurate demographic data on PIs and team members for competed missions, on grant proposers, as well as on the workforce across the four space mission-related science divisions of NASA SMD. Simultaneous to this study, another National Academies consensus study report committee was established and tasked with defining what data NASA needs to collect to assess the health and vitality of SMD's research communities. It is this committee's understanding that the Committee on Building a Foundation for Assessing the Health and Vitality of the Science Mission Directorate's Research Communities will specifically identify best practices for data collection in keeping with its charge. For this reason, this report will only briefly highlight strategies that have been employed at the federal level to maximize the benefit to society resulting from research funding.

### Evidence-Based Practices for Data Collection

NSF collects a variety of data about proposals, awards, proposers, and reviewers, and annually publishes a report, the Merit Review Digest,[3] that contains summary statistics of some of these data. Table 6.1, for example, shows the kind of information collected from PIs of proposals in 2019 in order to assess diversity of participation in NSF's research activities, and a full list of the comprehensive data included in the annual report is described in Annex 6.B. It is important to reiterate that while demographic information about proposers is based on self-reported data, and not all proposers choose to disclose this information, 71-78% of proposers for NSF funding elected to do so. Thus, this data collection strategy provides important information across virtually every dimension of the proposal review/award process, but more importantly, demonstrates transparency and visibility into investments for all stakeholders.

### Promising Practices for Data Collection

Although PI-led missions as we know them today began in earnest in the 1990s, information on PIs and team composition began to be collected for the first time by NASA SMD more than a decade later. The contents of this database have been enhanced with the development and implementation of NASA's Solicitation and Proposal Integrated Review and Evaluation System (NSPIRES), but the fact that the provision of demographic information is voluntary is frequently cited by the agency as a major barrier for collecting more comprehensive data on proposers. Other federal agencies (e.g., NIH and NSF) have also acknowledged this as a challenge, along with maintaining up to date demographic information on proposers as well as on the composition of the agency's respective workforces. Additionally, these agencies have implemented strategies to work around these challenges, such as obtaining demographic information by way of leveraging data generated by other organizations. In the past, NSF for example, has worked with the Higher Education Research Institute at the University of California, Los Angeles, to obtain demographic data on students and faculty in STEM fields to use in its National Science and Engineering Indicators biennial reports. Similarly, NIH has leveraged data from both the National Center for Education Statistics (NCES) and the American Association of Medical Colleges (AAMC) to track the distribution

[3] National Science Board, "NSF Merit Review Reports," https://www.nsf.gov/nsb/publications/pubmeritreview.jsp.

**TABLE 6.1** Participation in NSF's Research Activities by Members of Different Racial and Ethnic Groups

| | Hispanic | Non-Hispanic | Unknown | Total |
|---|---|---|---|---|
| American Indian or Native Alaskan | 27 | 40 | —[a] | —[b] |
| Asian | 20 | 7,533 | 590 | 8,143 |
| Black/African American | 23 | 626 | 21 | 670 |
| Native Hawaiian or Pacific Islander | —[a] | 29 | —[a] | —[b] |
| White | 795 | 15,023 | 986 | 16,804 |
| Multi-racial | 54 | 308 | 21 | 383 |
| Unknown | 334 | 1,040 | 6,135 | 7,509 |
| TOTAL | 1,253 | 24,599 | 7,761 | 33,613 |

NOTES: This table reflects the number of research proposals submitted by various racial and ethnic groups in fiscal year (FY) 2019. Demographic information about principal investigators of research proposals acted on in FY2019 is based on self-reported data.

[a] Number less than 10.

[b] Row sum not available because a cell includes a number less than 10.

SOURCE: Adopted from NSF's Merit Review Process: FY 2019 Digest.

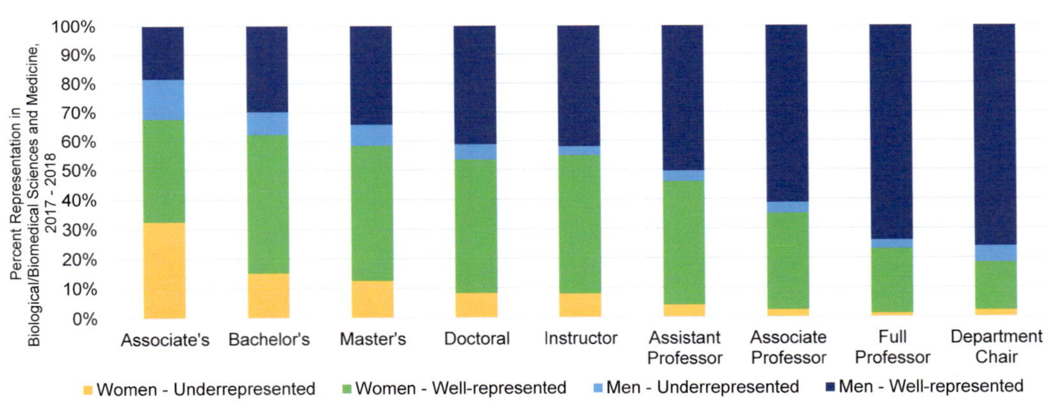


**FIGURE 6.1** Snapshot of the biological/biomedical workforce in 2017-2018, disaggregated by (binary) gender, highest degree earned, and status of representation in STEM. SOURCE: Marie Bernard, National Institutes of Health (produced from data from NCES and AAMC).

of scientists by gender, race and ethnicity through the pathway from undergraduate matriculation to the highest levels of academic medicine leadership (see Figure 6.1). In the physical sciences, we expect that such distributions will look different for women of all races and underrepresented men, and so this figure is simply meant to illustrate the data collection practices currently in place at NIH. Additionally, NIH tracks funding rates, as well as gender and career stage of all its PIs of Program Project Grant (P01) proposals across all institutes and centers.

**Finding:** NASA's Solicitation and Proposal Integrated Review and Evaluation System (NSPIRES) has enhanced the quality of its database of information on proposers, including mission PIs. Yet, the agency does not currently have much of the needed infrastructure, including adequate staffing, standardized data collection practices, monitoring and analytic systems, annual reporting capability, and external partnerships that other agencies like NSF and NIH have leveraged to overcome the challenge of tracking participation in S&E research activities via demographic information acquired on a voluntary basis.

**Conclusion 6-1:** Both the lack of a transparent, systematic process for assessing proposal reviews and the inadequate infrastructure for collecting demographic data on proposers severely limit NASA SMD's ability to accurately determine the relationship between the current mission proposal process and diversity in the pool of PIs. Furthermore, these insufficiencies result in limited ability to identify barriers in the mission proposal process and consequently the specific kinds of interventions that are needed to eliminate them and make progress toward the stated goal of effectively increasing DEIA in the leadership of competed missions.

## ACCOUNTABILITY AND PROMOTING DEIA IN S&E FIELDS

The NSF merit review process described earlier is overseen by the National Science Board (NSB), which establishes the policies of NSF within the framework of applicable national policies set forth by the president and Congress. In this capacity, NSB identifies issues that are critical to NSF's future, reviews NSF's strategic budget directions and the annual budget submission to the Office of Management and Budget, and approves new major programs and awards. Other agencies including NASA, NIH, DOE, and DoD also have oversight mechanisms in place, including but not limited to a "Committee of Visitors" for reviewing the management of different federal programs. DOE, for example, utilizes evaluation and management to monitor and assess its workforce training programs, and as part of the evaluation, a Committee of Visitors is used to provide the assessments of federal program management and the status of the programs relative to comparable programs supported by other agencies. Similarly, DoD's STEM portfolio activities and outcomes are guided by the DoD STEM Strategic Plan and overseen by an established STEM Advisory Council (SAC), but also managed and evaluated in accordance to DoDI 1025.11 (DoD 2020), a DoD-specific directive established in 2020. It is in the view of this committee that longstanding national advisory capabilities, comprised of nationally recognized experts in diverse disciplines, are essential for federal agencies to continue to help the nation leverage the talent of diverse communities to tackle today's highly complex challenges in S&E.

### Evidence-Based Practices for Accountability Through an Advisory Committee

In addition to NSB, the Committee on Equal Opportunities in Science and Engineering (CEOSE) is another congressionally mandated advisory committee that advises NSF on policies and programs concerning the implementation of the provisions of the Science and Engineering Equal Opportunities Act and other policies and activities to encourage full participation of women, minorities, and persons with disabilities in S&E fields. This committee holds meetings three times per year, and most recently gave focused attention to issues such as increasing accountability for broadening participation in S&E fields, addressing sexual harassment, promoting community engagement in science, understanding intersectionality, measuring broadening participation efforts, and promoting visibility and leadership in diversity, equity, and inclusion (DEI) initiatives and programs. Additionally, under the congressional mandate this committee is tasked with submitting a biennial report to the NSF Director and to Congress concerning:

- The state of participation of underrepresented groups in S&E fields;
- An assessment of NSF's policies and funding opportunities to broaden participation;
- Summary of all CEOSE activities during the 2-year reporting period and future plans for the next 2 years; and
- Recommendations to NSF for improving participation levels of underrepresented groups in S&E fields.

Most notably, recommendations from this committee have prompted (1) new and bolder agency-wide commitments to broadening participation, such as through the NSF Inclusion across the Nation of Communities of Learners of Underrepresented Discoverers in Engineering and Science (INCLUDES) initiative; (2) operationalized accountability for programs focused on broadening participation and for the agency itself; and (3) increased support for place-based implementation research projects that are grounded in and engage local communities, as well as increased attention to broadening the participation of diverse community members across its research and education portfolios through community-driven projects.[4]

### Promising Practices for Accountability Through Independent Advisory Bodies

Since 1977, the NASA Advisory Council (NAC) has served as the independent advisory committee through which NASA obtains advice, guidance, and recommendations on major program and policy issues before the agency. Currently, the NAC includes five committees that advise on matters related to aeronautics, human exploration and operations, science, STEM engagement, and technology, innovation, engineering, as well as a regulatory and policy committee. Upon further examination of the mission and scope of the NAC, the committee was able to obtain charter information for every subcommittee, except the STEM Engagement Committee, which appears to be the most recent committee established as part of the NAC in 2015. In 2022, DEIA is one of the NASA priorities to which the NAC will align their work. It remains to be seen how DEIA will be prioritized by the NAC, and thus what role it will play in providing guidance to NASA as it relates to SMD realizing its stated goals of broadening participation in PI-led missions in space science.

**Finding:** The NAC provides expert guidance, advice, and recommendations to NASA concerning a number of program and policy issues in S&E fields. However, there is a lack of evidence to substantiate that this committee is being appropriately leveraged by NASA to promote meaningful change in the area of broadening participation as is carried out by NSF's congressionally mandated CEOSE and NSB, NSF's oversight board.

**Conclusion 6-2:** The lack of a NAC committee focused on DEIA, and a respective committee chair serving on the NAC itself, misses a unique opportunity to both help set the tone at the top of the Agency and ensure ongoing and critical focus on shaping NASA SMD's broadening participation efforts in this area.

Alongside the efforts of standing advisory bodies, many federal agencies have enacted additional policies and procedures for promoting DEIA in S&E fields. In 2015, the Office of Science and Technology Policy (OSTP) and the Office of Personnel Management (OPM) established an interagency policy group, which is comprised of the Executive Office of the President and the relevant federal agencies. This working group was charged with cataloguing effective government-wide policies, practices, and strategies that have been used to mitigate the impact of implicit, institutional, and explicit bias in an effort to diversify their own STEM workforce (OSTP-OPM, 2016). Some identified policies and practices include evaluating recruitment, hiring, and promotion processes, conducting compliance reviews, and proactive use of DEI grants.

---

[4] See the Committee on Equal Opportunities in Science and Engineering (CEOSE) biennial reports to Congress for 2011-2012, 2015-2016, and 2017-2018, available at https://www.nsf.gov/od/oia/activities/ceose.

### Evidence-Based Practices for Accountability Through Evaluating Recruitment, Hiring, and Promotion Processes

Continual examination of recruitment, hiring, and promotion guidelines, as well as engagement strategies for current employees, is a common best practice used by federal agencies to promote DEIA in the workplace and foster equitable, safe, and respectful environments. NIH, for example, has done extensive work in this area. This includes, but is not limited to conducting periodic workplace climate surveys (NIH 2019a); analyzing recruitment, retention, and relocation incentives data to examine the racial and ethnic composition of employees receiving these incentives; reviewing the recipients of the NIH Directors Awards to determine whether racial and ethnic differences exist between groups in the nomination and selections rates for these honorary awards; and establishing expert committees to develop NIH-wide strategic planning for addressing any existing or potential barriers to DEIA in order to improve the culture of diversity and inclusion within the agency. Most recently, the agency has developed and published an *Ending Structural Racism* webpage[5] that offers a wealth of information, resources, and educational materials on understanding and combating systemic racism and having tough conversations about race in the work environment. NIH also regularly collects workforce demographics data on employees and publishes data on hires, separations, promotions, and awards in its biennial workforce demographic reports. A virtual dashboard[6] reflecting the latest demographic composition of the NIH total workforce has also been developed and these data are made available for all NIH employees, the entire biomedical research community, and the public.

### Promising Practices for Accountability Through Evaluating Recruitment, Hiring, and Promotion Processes

NASA has embraced some strategies to promote DEIA within the agency's internal workplace, which includes affirmative employment, conflict management training for employees, establishing an anti-harassment program, and a diversity and inclusion framework (NASA 2015b). As part of the diversity and inclusion framework approach, the agency has deployed comprehensive diversity and inclusion surveys to all employees for a more in-depth measurement of its diversity and inclusion efforts and to discover potential areas of bias or perceptions of exclusion that might exist in the workplace. NASA first deployed this survey in 2010 and repeated it in 2014. Since 2014, NASA has relied upon the New Inclusion Quotient of the annual Office of Personnel Management Federal Employee Viewpoint Survey to assess diversity and inclusion in the workplace.[7]

### Evidence-Based Practices for Accountability Through Compliance Reviews

Federal agency compliance reviews are often used to identify and examine any forms of bias (i.e., implicit, institutional, or explicit), discrimination, and/or harassment at the time of review of grantees and to provide specific recommendations for institutional leadership to reduce their impact. Many organizations and institutions who are recipients of federal dollars are obligated to comply with several regulations prohibiting discrimination of any kind. Of the list of regulations, the most germane to the scope of this report are Title VI of the Civil Rights Act of 1964,[8] which prohibits discrimination on the basis of race, color and national origin in federally assisted programs, and Title IX of the Education Amendments Act of 1972,[9] which prohibits sex discrimination in federally assisted educational programs. Upon further examination of the operational role of compliance reviews in the context of federal funding, we found limited, available data associated with the Title VI of the Civil Rights Act of 1964; therefore, we focus our discussion on Title IX compliance reviews.

---

[5] See the NIH "Ending Structural Racism" website at http://nih.gov/ending-structural-racism.

[6] See the NIH "Workforce Profile Dashboard" website at https://www.edi.nih.gov/people/resources/advancing-racial-equity/nih-workforce-profile-fy21q02.

[7] See the "Federal Employee Viewpoint Survey" at https://www.opm.gov/fevs.

[8] Title VI, Civil Rights Act of 1964, 42 U.S.C. § 2000d et seq: https://www.justice.gov/crt/fcs/TitleVI.

[9] Title IX of the Education Amendments of 1972, 20 U.S.C. § 1681 et seq: https://www.justice.gov/crt/title-ix-education-amendments-1972.

NIH is one of many federal agencies who have policies in place for addressing compliance with all applicable terms and conditions, policies, and requirements for institutions and individuals supported on federally funded awards, and in accordance with federal civil rights protections, which includes Title IX.[10] Specifically at NIH, pre-award assessments are conducted to verify that every applicant organization is eligible for funding on the basis of their status within multiple federal-wide systems used for management and oversight of federal funding recipients, and depending on specific circumstances, NIH reports instances of non-compliance into these systems. Post-award, recipient institutions are obligated to promptly report any administrative action taken, such as imposing leave or terminating employment, that affects the ability of senior/key personnel to continue on an NIH grant award, or otherwise constitutes a change in their status. A detailed description of NIH's standard procedures for handling allegations of harassment involving award recipients is publicly available.[11] Additionally, specific resources on preventing and addressing harassment and inappropriate conduct have been developed and are publicly available to the entire biomedical research community, including NIH employees. To document the impacts of these efforts, NIH regularly collects complaints-related data, and publishes these data as part of their annual Notification and Federal Employee Anti-discrimination and Retaliation (No FEAR) Act data reports submitted to Congress.

### Promising Practices for Accountability Through Compliance Reviews

In the case of NSF, the Office of Equity and Civil Rights (OECR) is charged with conducting Title IX compliance reviews, and selects sites for review in coordination with other agencies to avoid duplicative efforts (NSF 2019). To date, 11 reviews have been conducted and while there is no established process in place at NSF to specifically determine the effectiveness of Title IX reviews, the committee learned that current efforts in the area of compliance do include designing metrics for measuring the impact of the new term and condition (T&C) (NSF 2021a) established to address sexual harassment, other forms of harassment, or sexual assault in the context of NSF-funded activities and within the agency.

The committee also heard testimony from a former Associate Administrator of the Office of Diversity and Equal Opportunity (ODEO) at NASA about the nature of its Title IX compliance reviews, and learned that the agency conducts two reviews per year. To date, more than 20 reviews have been conducted, and identified areas of concern are documenting during these reviews, and a final report outlining corrective actions and recommendations are subsequently provided to institutions by the ODEO. However, similar to the NSF, no established metrics or processes have been established to determine the effectiveness of this form of accountability to date.

### Emerging Practices for Accountability Through DEI Grants

Proactive use of DEI grants afford programmatic support for developing and employing policies, practices, training materials, and recruitment and retention strategies designed to mitigate any bias within federally funded institutions of higher education. For example, beginning in 2017, the NSF Directorate for Computer and Information Science and Engineering (CISE) specifically enhanced support for PIs to learn about barriers to broadening participation in computing (BPC), with the ultimate goal of requiring PIs to include meaningful BPC plans in future proposal submissions. Similarly, in 2020 NIH put out a request for applications related to faculty institutional recruitment, as a mechanism for sustainable transformation at NIH-funded extramural institutions (RFA-RM-20-22). Specifically, this grant affords support for coordinating, collecting and analyzing data on institutional culture, diversity and inclusion, with the ultimate goal of recruiting and retaining a diverse group of early-career faculty who are competitive for an advertised research tenure-track or equivalent faculty position and who have demonstrated strong commitment to promoting diversity and inclusive excellence. The committee found no evidence of proactive use of DEI grants by NASA to date.

---

[10] NIH Grant Policy Statement: https://grants.nih.gov/grants/policy/nihgps/HTML5/section_4/4.1.2_civil_rights_protections.htm#Educatio; NIH Actions and Oversight to Address Harassment: https://grants.nih.gov/grants/policy/harassment/actions-oversight.htm.

[11] NIH Process for Handling Allegations of Harassment on an NIH-Funded Project at a Recipient Institution: https://grants.nih.gov/grants/policy/harassment/actions-oversight/allegation-process.htm.

**Finding:** With consideration for the suite of evidence-based, promising, and emerging government-wide practices that have been implemented, NASA has some analogous, albeit less developed, accountability mechanisms in place for promoting DEIA compared to other agencies like NSF and NIH. But, competed missions do not show robust evidence of positive outcomes across Divisions that advance these DEIA efforts. Additionally, there appear to be no established processes and metrics for evaluating the effectiveness of the current practices in place and for measuring progress toward stated goals.

**Conclusion 6-3:** The promise of NASA's existing best practices to address DEIA, both within the agency and PI-led mission opportunities, is limited by the lack of defined metrics to evaluate effectiveness and track progress resulting from these interventions.

## PROMOTING DEIA THROUGH INVESTMENTS IN STEM PATHWAYS

As first described in Chapter 4, individual pathways from education to occupation within S&E fields are often complex, and may include multiple entry and exit points along trajectories over the course an individual's academic and working lives (Cannady et al. 2014). Interestingly, about 20% of the U.S. STEM workforce majored in a field other than STEM (Landivar 2013). Thus, in this report, we emphasize "STEM pathways" rather than a "STEM pipeline" to better reflect the non-linear nature of paths into STEM fields (NASEM 2016; NSTC 2021). It is in the view of this committee that building and sustaining education-to-career pathways that invite and prepare the next generation of diverse STEM professionals is also critical for the future of competed space missions. Therefore, NASA SMD has an important role to play in bolstering STEM pathways. In the final section of this chapter, we discuss best practices for advancing DEIA along STEM pathways based on our review of the research literature and a range of STEM-related programs, research, and activities in the federal landscape. It is important to note that significantly more initiatives have been implemented than can be covered in this report, and so the policies and programs highlighted below are representative examples of the respective best practice, but do not reflect an exhaustive list. Additionally, rather than prescribe specific interventions along STEM pathways that NASA SMD should develop and/or support, the committee felt that it would be more beneficial to outline key considerations for the Agency as it makes future investment decisions in light of its appropriate role in this area. As an organizing framework for this section, our discussion focuses on six evidence-based guiding principles for STEM pathway investments that emerged through the committee's review: (1) promote the development of STEM identities, (2) establish flexible and relevant STEM education-to-career pathways, (3) intentionally recruit from historically underrepresented groups, (4) provide access to diverse mentorship, (5) foster career-life balance, and (6) promote systemic change around DEI.

### Guiding Principle 1: Promote the Development of STEM Identities

Research on student persistence in STEM trajectories reveal that the development of a STEM identity is a leading important factor, especially for students from historically underrepresented groups (Calabrese et al. 2013; Vincent-Ruz and Schunn 2018; White et al. 2019). In fact, programs that afford opportunities for the development of student's science efficacy, motivation, and identity are found to be the most successful at promoting persistence (Estrada et al. 2016). Additionally, STEM identities are shaped by opportunities to meaningfully engage in the practice of these disciplines and subsequently transfer this knowledge into action on issues of interest (Tan et al. 2013). Therefore, authentic and culturally relevant STEM engagement and research experiences are also critical to promoting the development of STEM identities. The development of a STEM identity is also positively associated with the likelihood of pursuing a STEM career (Stets et al. 2016; Cheng et al. 2018). To this end, a number of federal investments in STEM education have revolved around attracting a broad and diverse set of students and sparking interest in STEM, as well as providing opportunities for K-16 students to engage in valuable and authentic STEM research.

Both NASA's OSTEM and the Science Activation Program (SciAct) within NASA SMD have deployed a number of initiatives to inspire and motivate learners through exposure to NASA's various assets (e.g., state-of-the-art facilities; cutting-edge technology, missions and associated data and images). Specifically, NASA's Next Gen STEM efforts are mission-driven activities and have yielded a suite of evidence-based products. These facilitate opportunities for students to learn about new science discoveries and to acquire new skills in STEM. Additionally, the NASA Sounding Rocket RockOn! and RockSat-C Space Grant programs support experiences for middle school, high school and university students to design and build small science instruments for suborbital space flights in a 1-week workshop setting. Participants also gain experience in analyzing preliminary data from these flights, and can also take advantage of longer-term opportunities to deepen their skills in developing a scientific payload for flight throughout a school year. DoD offers a suite of activities for K-20 students, including but not limited to STEM camps whereby students connect with various STEM professionals around the country through STEM-focused lessons and talks, project-based competitions, student research opportunities, and paid apprenticeships and internships at DoD laboratories. Lastly, the annual Geostationary Operational Environmental Satellites (GOES-16/17) Science Fair, hosted by NOAA, affords opportunities for students in grade 6 through grades 13/14 (community college or university) to investigate weather and natural hazards using NOAA's satellite data. Participants also submit a project individually or as part of a small team to the science fair describing their findings.

### Guiding Principle 2: Establish Flexible and Relevant STEM Education-to-Career Pathways

Sustaining STEM identities once they have been developed is also critically important. Research shows that STEM interest and identity developed at one point in time could potentially be lost as individuals move toward careers due to a variety of factors (e.g., lack of mentorship, limited access to authentic research experiences, and mismatches between racial, ethnic, and gender identities and the perceived meanings of science and a scientist) (see Chapter 4 of this report) (Carlone 2004; Gazley et al. 2014; Remich et al. 2016). Supporting and cultivating STEM identities is also especially crucial for the retention of historically underrepresented groups in STEM. In the life sciences, self-efficacy in science research, greater intentions to pursue a STEM PhD, and higher scholarly productivity were all found to be associated with intentional efforts to sustain the scientific identities of students from groups underrepresented in STEM (Aikens et al. 2017). Opportunities that support gains in the ability to think and work like a scientist were also found to positively influence persistence in STEM, as well as serve as indicators of student growth as scientists and possible predictors of students' pursuit of science research–related careers (Estrada et al. 2011; Aikens et al. 2017).

Over the last two decades, some government-funded interventions to broaden participation in STEM have evolved to focus on establishing flexible and relevant STEM pathways that connect education and workforce opportunities, and accommodate students' diverse interests and backgrounds. For example, DOE supports tailored training opportunities specific to the fields and disciplines that are not adequately covered in university curriculum (e.g., high performance computing, accelerator science and technology, nuclear physics, nuclear chemistry, and isotope R&D). Additionally, the DOE Office of Science (SC) is starting a new initiative in FY2022, Reaching a New Energy Sciences Workforce (RENEW), that will leverage SC's world-unique national laboratories, user facilities, and other research infrastructures to provide undergraduate and graduate training opportunities for students and academic institutions not currently well represented in the U.S. science and technology ecosystem. DoD's Manufacturing Engineering Education Program expands opportunities for youth all the way up to adults already in the workforce to acquire deep skills in cutting-edge manufacturing techniques such as additive manufacturing and robotics, as well as shop floor skills like welding and machining. More broadly, DoD provides tuition support, stipends, and internships for undergraduate, master's, and doctoral students through its SMART Scholarships, which culminate in guaranteed employment at the agency.

The NASA Community College Aerospace Scholars (NCAS) Program consists of a 5-week online course led by engineers at different NASA centers and a 4-day engineering design workshop at a NASA center. Through this program, community college STEM students are also encouraged to participate in other competitive NASA projects (e.g., robotics competition) and internships, and to pursue a 4-year degree or career in a STEM field. Overall, the program aims to help students make the connection between a STEM degree and career opportunities at NASA. NOAA's José E. Serrano Educational Partnership Program with Minority Serving Institutions (EPP/MSI) Graduate Fellowship Program is a new program that provides professional development and 1 year of experiential training at a NOAA facility to students currently pursuing a graduate degree in a NOAA mission-related field specifically at an EPP/MSI Cooperative Science Center. Furthermore, the long-term goal of this program is to build out a workforce pathway for future employment at NOAA.

### Guiding Principle 3: Intentionally Recruit from Historically Underrepresented Groups

A critical driver to ensuring that the provision of flexible and relevant STEM pathways actually yield an inclusive STEM workforce is intentionality—a calculated and coordinated method of engagement used by institutions, agencies, organizations and the private sector to effectively meet the needs of a designated population (NASEM 2019a). As mentioned in Chapter 1 of this report, the inevitable changing demographics in the United States confer numerous opportunities for building a future, diverse STEM talent pool. Accordingly, a number of federal agencies have rolled out strategies to facilitate targeted recruitment of individuals from historically underrepresented groups at all levels including but not limited to actively engaging with MSIs that intentionally educate and serve the professional needs of these populations (NSTC 2021).

The Historically Black Colleges and Universities/Minority Institutions (HBCU/MI) Summer Research Program is a research experience for undergraduate and graduate students at DoD-supported HBCU/MI aimed at bridging classroom and real-world experiences in STEM disciplines, with a goal of increasing the number of individuals from historically underrepresented groups who choose careers in S&E fields. NSF has committed investments toward broadening participation in STEM, especially through its Inclusion across the Nation of Communities of Learners of Underrepresented Discoverers in Engineering and Science (NSF INCLUDES) initiative, which catalyzes the building of collaborative infrastructure for proactively seeking and effectively developing STEM talent at scale within populations that are historically underrepresented in STEM. NASA has recently joined the national network of partners for NSF INCLUDES and will specifically direct its efforts toward diversifying the engineering workforce,[12] a field that is indispensable to NASA's capacity for space exploration and discovery.

Within the past 5 years, NSF has also demonstrated more explicit commitments to building the capacity and research infrastructure at MSIs through the establishment of programs such as the Historically Black Colleges and Universities Excellence in Research (HBCU-EiR) program, the Tribal College & University (TCU) Enterprise Advancement Centers, and the Improving Undergraduate STEM Education: Hispanic-Serving Institutions (HSIs) program. NIGMS has implemented a suite of programs that collectively reflect targeted and intentional recruitment and development of diverse talent along a continuum of academic milestones (see Box 6.1).

Cultivating mutually beneficial partnerships and collaborations with MSIs is another facilitator of the recruitment and retention of individuals from historically underrepresented groups in STEM, and the NSF Partnerships for Research and Education and Materials (PREM) Program is one example of this approach. Specifically, PREM partnerships between MSIs and Division of Materials Research (DMR)-supported centers/facilities support recruitment, retention, degree attainment and pathways to careers for groups most underrepresented in materials science. Additionally, these partnerships are aimed at enhancing research productivity, programming, and infrastructure at partner institutions. Likewise, the NSF Louis Stokes Alliances for Minority Participation (LSAMP) program supports the recruitment and retention of students from underrepresented populations in STEM, primarily African Americans, Hispanics, American Indians, Native Hawaiians, and Pacific Islanders, through funding to alliances of institutions of higher education for the design, implementation, and study of STEM pathway-strengthening initiatives, such as bridge-to-doctorate programs and STEM Pathways projects. Similarly, NOAA's Educational Partnership Program with Minority Serving Institutions has demonstrated increases in retention and achievement of supported scholars from historically underrepresented groups in STEM. There are also key facilitators of the success of these partnerships including collaborative planning and program design with partner institutions, having the partnerships led by and the impacts mostly realized by MSIs, and a commitment by NOAA leadership to support the program. The NASA HBCU/MSI Technology Infusion Road Tour facilitates opportunities for MSIs to collaborate with NASA and its partners (e.g., Boeing and Lockheed Martin) in the context of a multi-day forum

[12] NASA and NSF INCLUDES Building MSI-Led Coalitions to Strengthen Broadening Participation in Engineering: https://www.nasa.gov/stem/murep/includes.html.

**BOX 6.1**
**Federal Investments in Recruiting and Retaining Diverse STEM Talent Along Academic Milestones: Examples from NIGMS**

**Undergraduate Level**

- **Building Infrastructure Leading to Diversity (BUILD)**—This initiative provides institutional-level support for innovative approaches to engaging and retaining students from diverse backgrounds in biomedical research, with the ultimate goal of positioning students on pathways to becoming future contributors to the biomedical research enterprise. BUILD institutions serve geographically and racially diverse population, and include HBCUs, HSIs, and Asian American/Native American/Pacific Islander-serving institutions (AANAPISIs).
- **Maximizing Access to Research Centers (MARC)**—This program supports the development and implementation of effective, evidence-informed approaches to biomedical research training and mentoring that yields a diverse pool of undergraduate STEM degree holders who go on to complete biomedical, research-focused higher degree programs (e.g., PhD or MD/PhD).

**Graduate Level**

- **Institutional Development Award (IdeA) Networks of Biomedical Research Excellence (INBRE)**—This initiative fosters development, coordination and sharing of research resources that lead to expanded research opportunities and an increase in the number of competitive investigators at academic institutions, who can attract talented students to bolster the science and technology workforce in IDeA-eligible states (states with historically low levels of NIH funding).
- **Native American Research Centers for Health (NARCH)**—These centers afford research opportunities and career enrichment aligned with health priorities established by American Indian/Alaska Native (AI/AN) tribes or tribally based organizations. Additionally, NARCH facilitate biomedical research careers related to AI/AN health and local and regional employment for American Indians and Alaska Natives.

**Post-Graduate and Faculty Level**

- **Maximizing Opportunities for Scientific and Academic Independent Careers (MOSAIC)**—This program facilitates the transition of promising postdoctoral researchers from groups underrepresented in the biomedical research workforce, into independent, tenure-track or equivalent research-intensive faculty positions. The related, MOSAIC Institutionally-Focused Research Education Award, provides support to scientific societies to equip MOSAIC scholars with professional skills, mentoring, and professional networks to support this transition.
- **Institutional Research and Academic Career Development Awards (IRACDA)**—This program supports the development of a diverse group of highly trained scientists to address the national biomedical research needs. It also promotes consortia between research-intensive institutions (RII) and partner institutions with a historical mission and a demonstrated commitment to providing training, education and support to students from groups underrepresented in the biomedical research enterprise.

SOURCE: National Institute of General Medical Sciences.

hosted on an MSI campus. Through this effort, MSIs are better positioned to acquire funding through more lucrative federal contracts, which in turn, enhances research capacity and competitiveness at these institutions (NASEM 2019a; NSTC 2021).

### Guiding Principle 4: Provide Access to Diverse Mentorship

Developing and retaining promising personnel through appropriate mentorship—including sponsorship, peer mentorship, and tiered mentorship—is also essential for advancing DEIA in STEM (NASEM 2019a; NSTC 2021). Effective mentorship affords individuals from underrepresented groups the potential to see themselves through the eyes of an influential role model (NASEM 2019b), and is closely tied to their development of a sense of belonging in STEM contexts (Byars-Winston et al. 2015). Furthermore, mentorship has been established as a high-impact and effective practice for promoting long-term student success in STEM fields (e.g., increased likelihood to pursue doctorates in STEM fields) (Kuh 2008; Hubbard and Stage 2010; NASEM, 2019a). Alternatively, lack of mentorship (formal or informal) or an inadequate support system can exacerbate feelings of isolation and invisibility in individuals from underrepresented groups, and as a result, negatively impact their persistence and success in STEM (Graham et al. 2013; NASEM 2019b).

Given the impacts of mentoring in producing a more diverse workforce and innovative research outcomes, a number of federal initiatives that leverage the benefits of mentoring in order to diversify respective STEM workforces have been employed. The NIH Distinguished Scholars Program, for example, aims to build a more inclusive community of NIH PIs through mentorship and other professional development activities that foster research and career success to investigators from diverse backgrounds. Similarly, the NIH National Research Mentoring Network (NRMN) is a nationwide consortium designed to enhance mentoring and career development of individuals from diverse backgrounds, including those from groups underrepresented in the biomedical research workforce. The NSF Research and Mentoring for Postbaccalaureates in Biological Sciences (RaMP) supports full-time research, mentoring, and training for baccalaureate graduates who have had few or no research or training opportunities during college, as a way to remedy the loss of diverse talent at junctions between the attainment of a STEM undergraduate degree and entry into a STEM career pathway. Additionally, the NSF Research Experience and Mentoring (REM) program provides authentic research experiences in STEM fields and ongoing mentorship and professional development to high school and undergraduate students, K-12 teachers, faculty, and veterans.

### Guiding Principle 5: Foster Career-Life Balance

As discussed earlier in this chapter, workplace climate can either be a facilitator or a significant barrier to realizing DEIA in STEM, as well as the well-being of all types of workers (Cech and O'Connor 2018; López et al. 2018; Cech and Blair-Loy 2019). Research shows that female employees are more impacted by work-life imbalance compared to male employees, but workplace flexibility bias—a sense that one will face career consequences for making work schedule adjustments for family or personal reasons—can negatively impact the well-being of all types of workers (Ong et al. 2011; Kachchaf et al. 2015; Cech and O'Connor 2018; López et al. 2018; Cech and Blair-Loy 2019).

Marriage and childbirth are key factors associated with the departure of women from the STEM workforce between PhD attainment and achieving tenure (NSF 2011). Consequently, some federal policies and practices have been implemented to minimize institutional barriers that can undermine performance and promotion due to parental status and other family responsibilities. For example, the NIH Research Supplements to Promote Reentry and Reintegration into Health-Related Research Careers support full- or part-time research by women or men returning to the scientific workforce after an interruption of at least 6 months for family responsibilities or other qualifying circumstances. The supplements also afford time for re-entering scientists to bring existing research skills and knowledge up-to-date, so that by the end of the supplement period, the scientist will be prepared to apply for other types of competitive independent research funding for continual support throughout their careers. NSF has also adopted similar policies and its Career-Life Balance Initiative was established to bolster the development of the STEM workforce through increasing the placement, advancement, and retention of women in STEM disciplines, particularly women seeking tenure in academia. Relatedly, Career-Life Balance supplements specifically support the expansion of eligible NSF awards to accommodate additional project personnel to sustain research while a researcher (e.g., graduate student, postdoctoral fellow, early-career scientist) supported by NSF on the award is on a leave of absence for dependent care responsibilities or other direct family considerations.

### Guiding Principle 6: Promote Systemic Change Around DEI

Wide-spread adoption of evidence-based practices is needed to achieve lasting progress on DEI in STEM fields, as well as systemic change within contexts where these practices are operationalized. To this end, some federal agencies have devoted resources toward preparing and empowering leaders and champions of diversity in order to transform the cultures of STEM fields. For example, in the geosciences, which has experienced little to no progress on increasing diversity in over 40 years (Bernard and Cooperdock 2018), targeted efforts have been put in place to build capacity for broadening participation. Specifically, as part of the NSF Geosciences Opportunity for Leadership in Diversity (GOLD) Program**,** geoscientists with interests in broadening participation receive professional development and training from experts in the areas of broadening participation, behavioral change, social psychology, institutional change management, leadership development and pedagogies in professional development, which effectively prepare them to be change agents and diversity leaders in the geosciences. The NSF ADVANCE: Organizational Change for Gender Equity in STEM Academic Professions supports systemic change and greater implementation of evidence-based strategies that facilitate equity and inclusion, while mitigating inequities in STEM academic professions. Additionally, realizing gender equity for STEM faculty in institutions of higher education is the preeminent program goal.

Similarly, the NIH Research on Interventions that Promote the Careers of Individuals in the Biomedical Research Enterprise Program strives to encourage the use of evidence-informed practices to enhance interest, motivation, persistence, and preparedness for careers in the biomedical research workforce. Moreover, addressing the impacts of structural racism, discrimination, and harassment on biomedical career progression are additional stated goals to change the culture of the biomedical research enterprise and ensure that future generations of biomedical researchers emerge from a diverse pool of talented individuals. Additionally, NIH's Training Modules for Creating Safe, Inclusive, and Supportive Research Environments are specifically designed to catalyze cultural change in scientific, clinical, and research training environments, such that they are free from harassment, intimidation and discrimination, and thus are optimized for productive learning and research. These training modules are geared toward a wide range of scientific disciplines and career stages, reflecting the goals of broader dissemination and impact. Finally, NIH established the NIH Equity Committee (NEC) in 2017 as part of its efforts to facilitate institutional transformation and culture change at both NIH and in the extramural research community. This committee is comprised of internal NIH staff and systematically reviews and assesses annual metrics on (1) the pool of tenured and tenure-track investigators in the biomedical workforce as it relates to equity in salaries and hiring and promotion policies and procedures; (2) equity in review practices (Board of Scientific Counselors and ad hoc reviewers, promotion, and tenure committees); (3) efforts to redress inequities; and (4) efforts to promote DEI.

Indeed, federal agencies have a non-trivial role in supporting diverse and inclusive education-career STEM pathways as demonstrated by the number of federal investments targeted at multiple points along these pathways that have been employed. Figure 6.2 illustrates where along education-career pathways select federal agencies have made sustained, intentional investments at any funding level toward advancing DEIA in STEM fields. This snapshot also highlights where purposeful investments are still needed at the federal level more broadly, and more specifically where there are opportunities for NASA to bolster its investments in order to leverage the potential of PI-led missions in space sciences. As described in Chapter 4 of this report, the first major drop in the participation of women (aggregate) in physics occurs during the high school to college transition, the first major drop in the participation of underrepresented groups (aggregate) in physics and astronomy occurs during the post-secondary stage, and attrition within both of these subgroups continues into the workforce stages. Thus, investments targeting multiple junctures along STEM pathways that could synergistically improve recruitment and retention in the Earth and space sciences may be needed.

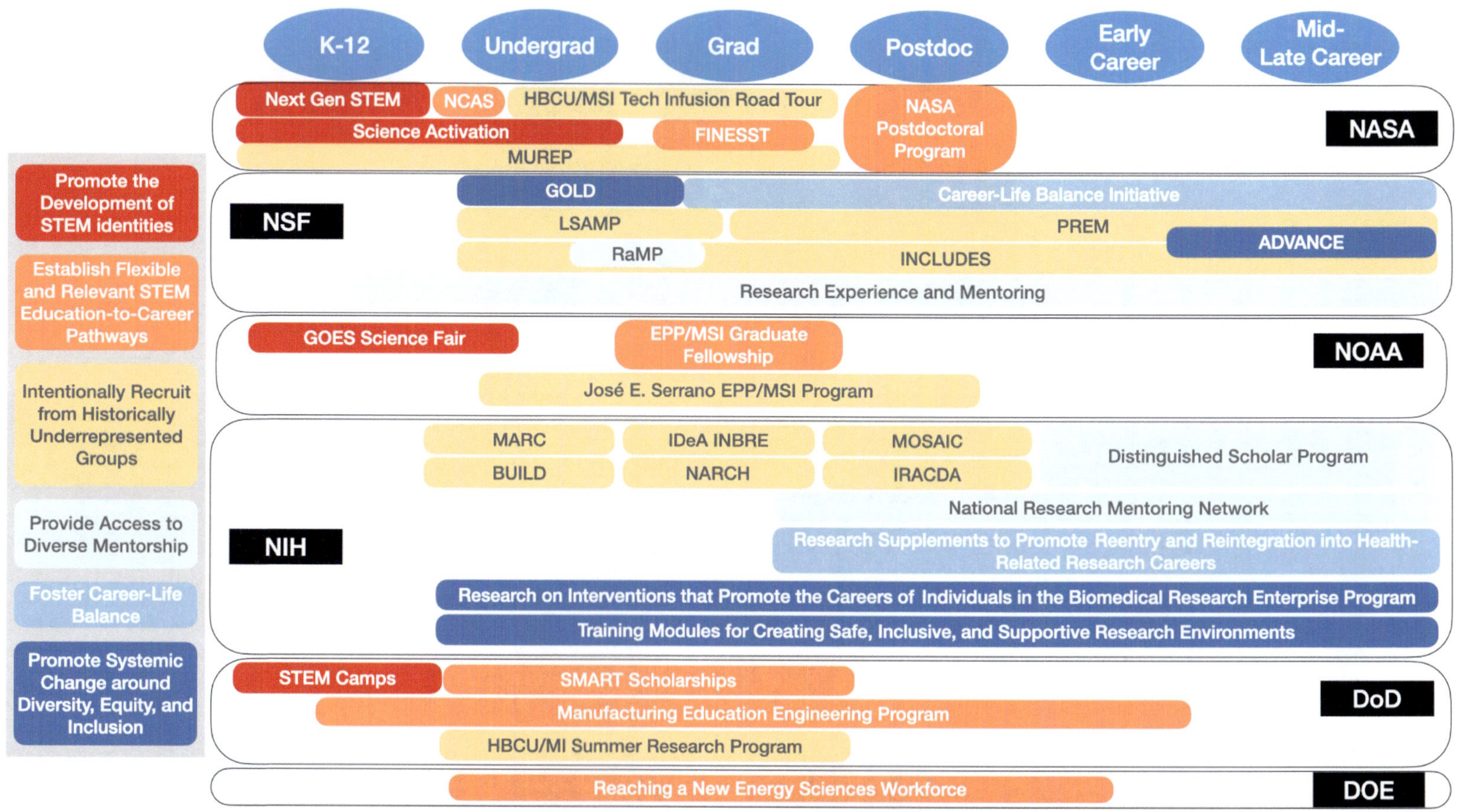


**FIGURE 6.2** Snapshot of select federal investments to advance DEIA along education-to-career pathways into STEM fields. For simplicity, each example is organized according to only one corresponding goal (*far left*), but it is important to note that more than one of these goals may be prioritized in a given intervention. The committee also acknowledges that education-to-career STEM pathways are not universally linear and so the respective placement of the different stages of education-to-career pathways (*top*) is meant to only facilitate the distribution of investments at the intersection of goals and target populations.

**Finding:** Current NASA investments in STEM education and STEM-related training programs and activities in the United States appear to be concentrated during the K-graduate years of STEM pathways, with very little intentional interventions employed during the post-PhD stage.

**Conclusion 6-4:** NASA's direct influence on the growth and development of a more diverse and inclusive Earth and space sciences workforce is potentially minimal due to limited investments in the post-PhD career stage. The lack of consistent and intentional investments post-PhD, may in part contribute to the underrepresentation of women of all races and men of color in the pool of leaders for competed space missions.

## ANNEX 6.A

### Details of SALMON and Program Element Appendix

SALMON are one solicitation mechanism used by NASA for announcements of opportunity (AOs). For example, the NASA AO, entitled "Third Stand Alone Missions of Opportunity Notice (SALMON-3)," provides a solicitation and procurement base for opportunities for modest investigations (missions of opportunities [MOs]) requiring spaceflight that advance the high priority science, exploration, and technology objectives of NASA's Science Mission Directorate (SMD), Human Exploration and Operations Mission Directorate (HEOMD), and Space Technology Mission Directorate (STMD). The SALMON-3 AO does not, in and of itself, solicit proposals. The actual solicitation is enabled by a Program Element Appendix (PEA) that is appended to the SALMON-3 AO. The AO provides the standard requirements for all SALMON-3 solicitations and specific requirements that may only apply to particular types of MOs. The PEA will call out the SALMON-3 specific requirements that apply and any additional program requirements for the specific solicitation and proposal opportunity. Program specific requirements spelled out in the PEA include the scope of the solicitation, the available funding, the proposal due date, and other program specific requirements as well as deviations or exceptions from SALMON-3 standard requirements (e.g., they may be tailored for two-step evaluations). The mission proposal opportunities are not on a fixed schedule, but a planning document that extends out a number years is available, such as https://soma.larc.nasa.gov/StandardAO/doc_files/Planning%20List%20for%20SMD%20Solicitations%2020210308-for-publication.docx.

## ANNEX 6.B

### Data Collection and Annual Reporting on Federal Awards at NSF

The following list, from the table of contents for the National Science Foundation (NSF 2020), provides a comprehensive list of the data included in NSF's Merit Review Process Annual Report:

III. Proposals and Awards
  A. Data on Research Grants
    A1. Research Proposal, Award, and Funding Rate Trends
    A2. Research Grant Size and Duration
    A3. Diversity of Participation
    A4. Number of Investigators per Research Project
    A5. Number of Research Grants per PI
    A6. Number of People Supported on Research Grants
    A7. Average Number of Months of Budgeted Salary Support for Single-PI & Multi-PI Research Grants
    A8. Principal Investigator Funding Rates
    A9. Early and Later Career PIs
  B. Competitive Proposals, Awards, and Proposal Funding Rates
  C. Diversity of Participation
  D. Types of Awards
  E. Awards by Sector and Type of Institution
  F. Time to Decision (Proposal Dwell Time)
  G. Mechanisms to Encourage Transformative Research
    G1. Small Grants for Exploratory Research (SGER), EArly-concept Grants for Exploratory Research (EAGER) and Grants for Rapid Response Research (RAPID)
    G2. Research Advanced by Interdisciplinary Science and Engineering (RAISE)
IV. The NSF Merit Review Process
  A. Merit Review Criteria
  B. Description of the Merit Review Process
  C. Program Officer Recommendations
  D. Review Information for Proposers and the Reconsideration Process
  E. Methods of External Review
  F. Data on Reviewers
  G. Reviewer Proposal Ratings and the Impact of Budget Constraints
  H. Program Officer Characteristics

# 7

# Paths Forward: Recommendations

The committee was tasked to address the issue of how to increase the diversity and inclusion in the leadership of competed space missions. The expertise of the committee spanned a wide range of experience from both scientists involved in National Aeronautics and Space Administration (NASA) missions as well as the social sciences. The committee approached the task through meetings with representatives of different groups and institutions, as well as contracted a qualitative study involving structured interviews with past principal investigators (PIs) of competed mission proposals.

This report begins with a description of the substantial, complicated process involved in developing, preparing and submitting a proposal for a competed mission. Some of these steps are formal, others are informal; some of the evaluation criteria involved in this process are explicit, others are tacit. In order to characterize the current state of diversity in the space sciences, the comparative analysis was made using the available demographic data on teams who have submitted competed mission proposals, on individuals who have submitted proposals to NASA's research and analysis programs, and on the overall space sciences profession. Examination of education-career pathways into the space sciences profession shows that the "pinch point" where the persistence of underrepresented minorities in the physical sciences significantly declines is at the early undergraduate stages, with severe under-representation persisting through all subsequent stages of the profession. Moreover, interviews with institutions and with individuals reveal that there are multiple barriers encountered along the path to mission PI-ship. Finally, the programs and processes that inspire, train, mentor the next generation of NASA space scientists—potentially leading to PI-ship of a competed mission—were compared with similar efforts at other agencies, leading to the identification of a number of best practices.

The committee approached the development of its recommendations with the goal of achieving two related but distinct outcomes to diversity, equity, inclusion, and accessibility (DEIA) for future NASA space missions. First, the diversity of the workforce leading into NASA mission leadership—not just within NASA, but also at colleges and universities, industry and other organizations—needs to significantly improve. This report provides quantitative analyses of the scale of the challenge and opportunity with respect to the makeup of the workforce (see Chapters 3 and 4). Second, in order to achieve—and importantly to sustain—the needed increases in NASA's workforce diversity, the system of processes and policies (e.g., proposal development, review and evaluation) that determine who gets access to NASA mission involvement and leadership opportunities need to be made more equitable, inclusive, and accessible (see Chapters 2 and 5). In its current form, the process for proposing a competed mission to NASA is closed, opaque, and exclusive. Processes that are closed/siloed/opaque present barriers, particularly disadvantaging underrepresented groups, and thereby hindering efforts to realize DEIA. Both the "D" and the "EIA" of DEIA are important outcomes, one in the form of the people who make up the NASA mission workforce and leadership, and the other in the form of the architecture of the system that includes, prepares and supports that workforce toward mission engagement and leadership. Furthermore, achieving diversity of the workforce requires sustained DEIA efforts. Some past efforts by NASA on DEIA have been defunded (see Chapter 4). Mechanisms for assessing the effectiveness of such programs are routinely applied at other agencies (see Chapter 6), which in part, allow them to sustain effective programs.

This chapter puts forward recommendations that appear based on the timeframe for their implementation—from near term to long term, and it is in the view of this committee that to sustain any specific actions taken by NASA Science Mission Directorate (SMD) to realize DEIA in competed missions, a cultural change—changed mindsets, policies, and practices across multiple levels—is needed across the organization along with collaboration across different divisions that usually operate in siloes. At the very highest level, the lack of a NASA Advisory Council (NAC) committee focused on DEIA misses a unique opportunity to both help set the tone at the top of the Agency as well as ensure ongoing and critical focus on shaping NASA SMD's broadening participation efforts in space missions.

**RECOMMENDATION 1: NASA should empanel an ongoing NASA Advisory Council (NAC) committee specifically focused on diversity, equity, inclusion, and accessibility (DEIA), whose committee chair serves directly on the NAC. This committee should have a broad charter and external world-class membership in this area to directly advise top NASA leadership and ensure an ongoing strong focus on NASA's broadening DEIA efforts.**

At the same time, the existence of a NAC committee specifically focused on DEIA does not eliminate the need for DEIA to also be a priority for the other six committees of the NAC. Therefore, as the NAC undergoes reconstitution, there is also an opportunity to articulate how each committee's focus on DEIA in ways that are most appropriate given its respective charter, aggregates toward meeting NASA's broadening participation objectives.

Since the competed missions are managed by four science divisions of SMD—Heliophysics, Astrophysics, Planetary Science, and Earth Science—the following recommendations are aimed at these divisions, and are gathered together under four general themes: (1) Proposal Process and Review: Enhancing Opportunities, Mitigating Barriers; (2) Data System: Collection, Analysis, and Reporting of Data on Proposal Teams, Proposer Pools, and Workforce; (3) Training and Mentoring Potential PIs; (4) Investment in Career Pathways for Underrepresented Groups. However, most of the recommendations could also apply to all areas that fall within NASA's sphere of influence, and if the suggested actions were carried out throughout the agency and organizations funded by NASA, then a more diverse workforce would drive U.S. endeavors in space exploration.

## THEME 1: PROPOSAL PROCESS AND REVIEW: ENHANCING OPPORTUNITIES, MITIGATING BARRIERS

NASA's competed space missions are selected from proposals that are submitted in response to a public announcement of opportunity (AO). The committee found many ways in which NASA could improve the proposal process to make it more accessible, inclusive, and equitable, from making the process of proposal development and submission more transparent to including DEIA as evaluation criteria to simplifying site visits to reducing bias in the review process. Moreover, there are valuable lessons to be learned from other federal agencies.

Before the AO is made public, the process of team formation and concept development requires significant resources from proposers and their institutions. These stages of the proposal process are often referred to as "the competition before the competition" which, is in part informal, idiosyncratic, organic, opaque and often personality-driven. Moreover, these informal and organic processes, influenced by who knows whom in the community, directly impact the diversity of the PI candidate pool.

For example, many proposals for competed missions are submitted by NASA centers. Thus, NASA is a leading force in the overall demographic makeup of the pool of mission PIs. Moreover, across the range of institutions submitting proposals, the lack of uniformity in internal resources available for preparation and production of mission concepts, full proposals, and site visits may result in a disadvantaged position for some institutions compared to others. For example, larger, well-resourced institutions often have teams with specific roles to support the proposal process, such as managers, administrative staff, financial advisers, expert engineers, internal reviewers, graphic artists, editors, etc. Likewise, in addition to such internal resources, NASA center teams likely have a greater understanding of the proposal process (e.g., reporting requirements, administrative processes, etc.), which is undoubtedly an important factor for developing high-quality mission proposals. The process of PI selection for NASA center-led competed missions could be much more transparent and committed to partnering with and supporting traditionally underrepresented PIs. A restructuring of this type would provide a roadmap for other institutions to follow in their own PI selection processes.

**RECOMMENDATION 2: NASA should work to make the pre-proposal "competition before the competition" process transparent and accessible. Additionally, NASA should use its own resources to expand support of pre-proposal and proposal efforts of diverse, external principal investigators through its field centers and encourage other institutions in the business of supporting and investing in Science Mission Directorate proposals and missions to do the same.**

One demanding component of the mission proposal process is the site visit where a NASA review team visits the proposing institution for a presentation of the proposal by the PI's team followed by questioning by the review team (see Chapter 2). Over time, such site visits have grown in scale in terms of required resources, time, and costs with the tendency to escalate into an elaborate event in execution (see the qualitative study in Appendix C). As the proposal process for competed missions has gotten more expensive and complex, it is likely that the cadre of institutions that can credibly lead them has gotten smaller.

**RECOMMENDATION 3: NASA should reconsider the requirements for site visits to emphasize the evaluation of technical and programmatic readiness, and eliminate any unnecessary elements. NASA should evaluate the benefit of providing uniform funding to each team that is preparing a site visit, disallowing supplemental funding and other contributions that may result in inequities across teams.**

The steps of the NASA AO process for competed mission proposals are governed by detailed regulations, but scrutiny for sources of bias throughout the implementation of the review process is warranted, and the process should also include DEIA as explicit criteria for evaluation. Furthermore, the fact that bias review and DEIA criteria are not currently used in a consistent and systematic fashion, results in limited ability to identify barriers in the mission proposal process, and consequently the specific kinds of interventions that are needed to eliminate them and make progress toward the stated goal of effectively increasing DEIA in the leadership of competed missions.

**RECOMMENDATION 4: NASA Science Mission Directorate (SMD) should develop and make public a systematic and transparent process to assess how the review of proposals submitted for research support is conducted.**

- **Moreover, NASA SMD should collaborate with experts to develop and employ an assessment of its mission proposal review process as well as involve experts on disparities in research funding experienced by historically marginalized groups. Such an assessment should also include gathering feedback from proposers.**
- **In the longer term, NASA Headquarters should develop a comprehensive assessment of its proposal review processes, not only with the principal investigator–led missions, but to be employed agency-wide.**
- **This analysis and assessment should include consideration of emergent bias-reducing practices (such as Dual-Anonymous Peer Review) and methods to track, identify, and reduce bias in the review and evaluation process.**

**RECOMMENDATION 5: In keeping step with its core values of diversity, equity, inclusion, and accessibility (DEIA), NASA Science Mission Directorate should:**

- **Require announcements of opportunity to include a description of how the dimensions of DEIA, including talent development and workforce diversity, are critical to NASA, and require proposals to include a plan for DEIA in the proposed missions describing how the proposed DEIA activities are key for mission success.**
- **Establish a separate, scorable evaluation criterion of the proposed DEIA plans as part of mission proposal review and provide training for reviewers to better equip them to appropriately evaluate proposals concerning DEIA dimensions.**
- **Engage with DEIA experts to implement the new requirements for mission proposals in ways that advance the overarching goal of broadening participation of underrepresented groups in missions.**
- **Establish monitoring and assessment processes to continually measure the impact of the new DEIA plan requirement on progress toward NASA's stated DEIA goals.**

**RECOMMENDATION 6: NASA should regularly monitor and assess adherence to the proposed diversity, equity, inclusion, and accessibility plans throughout the mission lifecycle and require up-to-date reporting on climate within mission teams in ways that go beyond compliance. For designing and interpreting climate assessments, NASA Science Mission Directorate should engage with content experts (e.g., survey design experts, social science scholars).**

Additionally, it is important to note that PIs and mission teams may need specific training and supports to develop feasible and appropriate plans that effectively work toward realizing DEIA beyond achieving greater numerical representation across different demographic groups. Therefore, NASA SMD may also want to consider offering workshops and training sessions for mission teams (proposing and existing) on creating and implementing effective DEIA plans, similar to those that focus on writing successful proposals. Such professional learning opportunities could be provided directly by NASA or in partnership with professional space sciences organizations. Similar training at proposing institutions where these plans will be executed may also be needed, given experiences of bias, discrimination, and exclusionary interpersonal interactions hinder the creation of an inclusive work climate. Thus, it may take coordinated efforts at multiple levels of the system to yield the organizational culture change that is needed to fully realize DEIA in the context of competed missions.

## THEME 2: DATA SYSTEM: COLLECTION, ANALYSIS, AND REPORTING OF DATA ON PROPOSAL TEAMS, PROPOSER POOLS, AND WORKFORCE

Inadequate data gathering, monitoring and reporting are key barriers to NASA's understanding of the demographics of their proposer pool, and to their ability to measure improvements. Specifically, the inadequate infrastructure for collecting demographic data on teams submitting mission proposals severely limits NASA SMD's ability to accurately determine the relationship between the current mission proposal process, the diversity in the pool of PIs, and the overall workforce in the space sciences. For example, further study is needed to evaluate to what extent variations over time and differences between SMD divisions in the proportion of women among mission PI proposers compared to the proportion of women among doctorate recipients in the physical sciences are due to the mismatch between the career stage distribution of women in these various populations and the current seniority expectation for mission PIs. The current data gathered by NASA on the demographics of proposal teams, for both competed missions and grants, is inadequate to inform NASA about the diversity in participation of Earth and space scientists from different demographic groups in research and activities funded by the Agency. Furthermore, promise of NASA's existing best practices to address DEIA, within the agency, across the space sciences workforce, and on PI-led mission teams, is limited by the lack of defined metrics to rigorously evaluate and track the progress resulting from these interventions.

**RECOMMENDATION 7: NASA Headquarters (HQ) should develop a systematic and transparent process that employs routine monitoring and tracking of proposal submissions and selections, and submit an annual report of these data to the diversity, equity, inclusion, and accessibility committee of the NASA Advisory Council as well as make the report publicly available. This report should include data on dimensions such as funding rates and diversity in team participation in principal investigator–led missions as well as research and analysis grants; but could also include data on other important dimensions of interest to the Agency. NASA HQ should seek professional statistical expertise to set in place the needed infrastructure to support robust data collection, monitoring, and reporting including, but not limited to, adequate staffing, data collection standards of practice, monitoring and analytic systems, annual reporting capability, and external partnerships, to overcome the challenges of tracking participation in NASA's Earth and space science activities.**

Proposals to NASA's SMD are submitted via an online system (NSPIRES) that requires the proposal team to fill in a Personal Profile that comprises answering some basic questions about gender, ethnicity, race, and disability/serious health conditions. Presentations from NASA's Office of the Chief Scientist and analysis of these Personal Profile data revealed that there are several users (~20%) who select Prefer Not to Answer in response to these questions, which limits the potential of these data to inform decision-making around broadening participation in NASA's Earth and space science activities. Moreover, NSPIRES users are not regularly requested to update their profile. Other government agencies (e.g., NSF) routinely gather such proposers demographic data and regularly report on their analyses. Most important is that experts from the social sciences are consulted on the nature of the questions as well as best practices for enhancing response rates.

**RECOMMENDATION 8: Working with experts in demographics data gathering and analysis, NASA should review, update and expand the NSPIRES Personal Profile questions and regularly encourage proposers to update their responses.**

The issue of lack of demographic data is not limited to NASA proposals. There is currently no strategy and methodology being employed to effectively gather demographic data on participation in the Earth and space sciences along the whole career pathway—from undergraduate and graduate programs to academic departments to the professional workforce. The American Institute of Physics (AIP) is an umbrella organization that encompasses organizations such as the American Physical Society, the American Geophysical Union, the American Astronomical Society, etc., and includes a Statistical Research Center with experience in surveying the workforce of the corresponding academic fields. While NSF and AIP regularly gather information about academic departments (such as numbers and types of bachelor's degrees, doctorate degrees, and faculty in different science fields), the workforce in the Earth and space sciences is usually categorized according to the type of object observed and therefore does not align directly with the disciplinary categories typically represented in such studies. Not only are the SMD divisions interdisciplinary in nature (e.g., Planetary Science comprises geology, biology, atmospheric science, space physics, etc.) but there is also considerable overlap in the science covered by different divisions. Currently, both the total size of the U.S. post-PhD workforce in each of the disciplines associated with the four divisions of SMD with competed missions, and the degree of the overlap between those disciplines, are very poorly quantified. The knowledge and experience of the AIP in working with professional organizations should be combined with rigorous statistical expertise in survey development and analysis. Workforce surveys have recently been included as part of decadal surveys, but to inform NASA and the space sciences community of changes in the workforce, such surveys need to be carried out and the results made public every 3-5 years.

**RECOMMENDATION 9: To regularly assess the state of the profession, NASA Science Mission Directorate should provide funding for professional organizations (e.g., American Institute of Physics, American Astronomical Society, American Geophysical Union, American Physical Society, etc.) to employ the necessary professional expertise in survey methodology and statistical analysis to systematically carry**

**out surveys of the workforce, within and across the four science divisions with competed missions, to inform NASA of the participation of different demographic groups as well as the barriers and opportunities for advancement along entire career pathways in the Earth and space sciences.**

Such data and analytics on participation in the Earth and space sciences, in the NASA-funded workforce, and on mission teams will enable NASA SMD to monitor trends and evolution over time, and inform decision-making around broadening participation with better clarity, for example, around "who" is in the profession, where barriers remain, and consequently what NASA-specific efforts can help to mitigate them. Moreover, these data could also be leveraged by other agencies that support research in the Earth and space sciences (e.g., NSF, the National Oceanic and Atmospheric Administration, the Department of Energy, and the Department of Defense) as well as by interagency groups, such as the Committee on Science, Technology, Engineering, and Math Education (Co-STEM), to yield better coordination of federal investments toward broadening participation in science, technology, engineering, and mathematics (STEM).

## THEME 3: TRAINING AND MENTORING POTENTIAL PIs

Preparation for competed mission leadership starts early in a space science career. Training and mentorship opportunities are valuable tools to develop a diverse pool of future mission leaders. The informal pre-proposal process presents critical challenges to the aspiring PI, and especially PIs from small and less-resourced institutions. A small number of individuals and institutions are effectively "gate keepers" for selection of mission PIs, and non-PhD granting academic institutions, Historically Black Colleges and Universities (HBCUs), and other Minority-Serving Institutions (MSIs), where a high concentration of woman physics faculty and Black and Latinx physics faculty are employed, experience inequitable access to opportunities that are helpful to scientists who go on to submit competed mission proposals. Proposal development, preparation, and submission—including site visits—are resource-intensive processes, and therefore diversifying the PI population will require investment in a range of institutions, but particularly in less-resourced MSIs. Furthermore, the collective evidence that the committee analyzed in its work suggests that the onus is not primarily on the individual when it comes to enhancing the likelihood of success as a future leader of a competed space mission given the range of structural, institutional/organizational and interpersonal factors that are major hindrances to assuming leadership positions on competed space missions. Therefore, most of the recommended actions within this theme are directed at the organizational/institutional level rather than the individual level.

**RECOMMENDATION 10: NASA should expand and increase the frequency of training programs that are aimed at encouraging women and historically minoritized communities to become more involved in mission leadership.**

**RECOMMENDATION 11: To engage and train diverse teams at all stages of professional talent development, NASA should offer mission-related research, mentorship, and training opportunities—ideally, integrated into actual NASA missions—through colleges/universities as well as NASA centers, that should start as early as first-year undergraduates and graduate students (e.g., internships), and extend to the ranks of postdocs (e.g., fellowships), and established scientists (e.g., participating scientists).**

Thus, not only does NASA need to make opportunities to lead missions more accessible and transparent, but also be directly involved in expanding the pool of potential PIs at various stages of professional development. Examples of experiences for mission-related engagement and training could include participation in mission summer schools organized for graduate students and junior post-docs, attendance at mission-related workshops at scientific conferences or at proposing institutions, and involvement in mission science team meetings.

**RECOMMENDATION 12: Principal investigator (PI)-led missions present opportunities for aspiring PIs to gain invaluable experiences. NASA should expand resources (e.g., instructional materials, seminars, workshops) for aspiring PIs to gain leadership experience and connect with individuals with mission experience for mentorship opportunities. This may include:**

- **Integrating aspiring PIs as mentees in roles on mission teams, including the higher leadership positions. This could be achieved by including developmental positions in all missions (i.e., competed, non-competed, and instrument investigations), which may require increasing the PI-managed costs.**
- **Encouraging aspiring PIs to pursue entry points to mission leadership, such as proposing to smaller, low-cost mission opportunities, (e.g., suborbital, SmallSats and CubeSats, instrument development, and hosted payloads).**
- **Expanding structured networking opportunities at relevant disciplinary conferences such as meet-and-greets where aspiring PIs can connect with collaborators and meet existing PIs, and participate in presentations and question and answer sessions led by NASA personnel about the proposal process.**

**RECOMMENDATION 13: NASA should evaluate the skills and expertise needed for success as a principal investigator (PI) beyond scientific competencies, including abilities leading and managing diverse, equitable, inclusive, and accessible teams. This more expansive set of competencies should be reflected in discussions about PI-ship in instructional materials and other outreach efforts.**

Mentoring and access to collaborative networks that include experienced PIs are formative in preparing those who submit competed space mission proposals. However, women and racially minoritized space scientists report less access to mentors and a lower quality of relationship with their doctoral advisors and senior colleagues. Inequitable access to high-quality mentoring relationships is a barrier to increasing diversity in the leadership of competed space missions. Thus, there is a need for all mentors, regardless of identity, to adopt inclusive mentoring practices (see Chapter 5), to ensure that the benefit of high-quality mentoring relationships is more widely experienced. Moreover, given the demographic realities within the Earth and space sciences, the current pool of mentors (i.e., PIs, Co-Investigators, and others with relevant mission experience) may also need explicit training around advising and supporting early-career scientists from underrepresented groups. Additionally, given many of the skills that are required to be a successful PI are not generally incorporated in academic education, developing such competencies may need to be a central part of mission-related mentorship and training.

## THEME 4: INVESTMENT IN CAREER PATHWAYS FOR UNDERREPRESENTED GROUPS

The very low overall retention (~11%) in space science disciplines during undergraduate training, and the accompanying racial/ethnic disparity, is currently a major "pinch point" that limits representation at all later stages of the education-career pathway and restricts the size and diversity of the pool of PhD scientists for future NASA mission leadership. Interventions to improve the diversity and inclusion of STEM students—the ultimate source of proposer pools—are needed at the high school and college levels, where studies show that much of the diversity in relevant fields is lost. NASA needs long-term, sustained investments in effective activities that inspire, educate, train and mentor, to ensure that the currently small pool of scientists of color have every opportunity to engage in NASA mission-related work and leadership. NASA is uniquely positioned to provide individuals with research experiences that are relevant to future mission roles, and there is arguably nothing more authentic than experiences linked to actual NASA missions.

Informed by the existing knowledge base and recognized exemplar interventions, several guiding principles were identified to inform NASA's investment in these career pathways (see Chapter 6). The committee found it more important to recommend an intentional strategy through guiding principles to inform NASA's future investments for bolstering STEM pathways rather than prescribing specific interventions. The guiding principles include (1) promoting the development of identities; (2) establishing flexible and relevant STEM education-to-career pathways;

(3) intentionally recruiting from historically underrepresented groups; (4) providing access to diverse mentorship; (5) fostering career-life balance; and (6) promoting systemic change around diversity, equity, and inclusion.

Historical underinvestment in the space science R&D infrastructure at Historically Black Colleges and Universities and other MSIs has in effect limited the capacity at these institutions to compete for a NASA space mission. Moreover, given the racialized patterns of where today's faculty are employed and where degree earners in SMD-related fields are educated, historical underinvestment leads to a disproportionately negative impact on faculty members of color in SMD-related fields and limits access to early-career mission experiences for students of color at MSIs. A previous NASA SMD-funded program, MUCERPI (Minority University and College Education and Research Partnership Initiative), did afford students with training in state-of-the-art research methods and preparation for future NASA leadership, specifically through partnerships with MSIs, but has been defunded. The Minority University Research and Education Project (MUREP) and MUREP Institutional Research Opportunity (MIRO) through NASA's Office of STEM Engagement (OSTEM) provide specific engagement to underrepresented and underserved communities in STEM, and develop the research capacity and infrastructure of MSIs in areas of strategic importance and value to NASA's mission and national priorities. Further partnership among NASA SMD, OSTEM, and MSIs, leveraging NASA's unique assets, and making NASA-led opportunities more perceptible to individuals and organizations would strengthen participation of underrepresented groups in missions. Likewise, NASA's current investments in STEM education and STEM-related training programs and activities in the United States appear to be concentrated during the K-graduate years of STEM pathways, with very little intentional interventions employed during the post-PhD stages where attrition continues to be observed. Thus, NASA's influence on the growth and development of a more diverse and inclusive space sciences workforce is potentially minimized by limited investments along the career path at the post PhD stage.

**RECOMMENDATION 14: In order to ensure a vibrant, next generation pool of excellent and diverse talent for leadership in competed space missions, NASA Science Mission Directorate, in collaboration with the Office of STEM Engagement, should provide consistent and adequate funding for science, technology, engineering, and mathematics initiatives that are explicitly centered on diversity, equity, inclusion, and accesibility, address recruitment and retention challenges in the Earth and space sciences, and support and expand opportunities for individuals from underrepresented groups. These investments should reflect a pathways approach spanning the academic and career continuum from post-secondary through post-PhD years in order to establish flexible and robust education-to-career trajectories into the Earth and space sciences workforce, and ultimately into principal investigator–led missions. A systematic process should also be in place to document measurable impacts of these investments.**

**RECOMMENDATION 15: Recognizing the critical role that Historically Black Colleges and Universities (HBCUs), Hispanic-Serving Institutions (HSIs), and other Minority Serving Institutions (MSIs) play in educating and employing women and racially minoritized populations in the Earth and space sciences workforce, NASA leadership, specifically the Science Mission Directorate (SMD) and the Office of STEM Engagement Associate Administrators, should charter a joint team to examine and strengthen the historic and current relationship between the two organizations with respect to investments in MSIs. NASA's investments should also redress the historical inequities in NASA supported research and training at these institutions. Specifically, NASA should:**

- **Reinvest in talent development programs in partnership with MSIs specifically related to NASA missions, such as the undergraduate-to-graduate "bridge" type programs previously supported by NASA SMD's Minority University and College Education and Research Partnership Initiative program.**
- **Further leverage NASA programmatic assets such as Minority University Research and Education Programs (MUREP), MUREP Institutional Research Opportunity, and Science Activation to advance broad access to all of its missions, and further enhance early preparation and research engagement of students and early-career researchers, including underrepresented communities.**

- **Provide funding to support mission-related work and activities (e.g., building and designing instruments for spaceflight, hosting science team meetings, etc.) as a means of enhancing research capacity at HBCUs, HSIs, and other MSIs.**

The committee acknowledges that to enact these proposed investments, NASA may need to specifically request funding for them and receive subsequent congressional approval. Nevertheless, it is in the view of this committee that such investments are needed to better position HBCUs, Hispanic-Serving Institutions, and other MSIs to lead missions as the primary institution and in turn help the Agency make significant progress in the area of DEIA. Therefore, these investments need to be considered as part of upcoming budget plans notwithstanding final budgetary decisions that may be beyond NASA's control. Additionally, the committee believes that joint SMD and OSTEM efforts and investments described in the above recommendations could be activities for the newly-constituted DEIA NAC committee to follow, advise, and measure their effectiveness.

Examination of the demographics of the leadership of NASA's competed missions clearly shows a serious paucity of diversity. Diversifying the full STEM workforce will take a major national effort. In the meantime, there are several actions that NASA could take, from the immediate clarification of the proposal process to training and mentoring students and early- to mid-career space scientists to intentional, longer-term investments in educational experiences that involve college students in space missions. By stepping up to such initiatives, NASA would not just be following the example of other agencies but instead would be leveraging the great excitement of space exploration to lead the way forward in diversifying the U.S. STEM workforce.

# References

AAUW (American Association of University Women). 2008. "The STEM Gap: Women and Girls in Science, Technology, Engineering and Mathematics." Washington, DC. https://www.aauw.org/resources/research/the-stem-gap.

Abrica, E.J., T.B. Lane, S. Zobac, and E. Collins. 2020. "Sense of Belonging and Community Building within a STEM Intervention Program: A Focus on Latino Male Undergraduates' Experiences." *Journal of Hispanic Higher Education* 21(2):228-242. https://doi.org/10.1177%2F1538192720974884.

Agresti, A. 2002. *Categorical Data Analysis*, 2nd ed. John Wiley & Sons, New York.

Aikens, M.L., M.M. Robertson, S. Sadselia, K. Watkins, M. Evans, C. R. Runyon, L.T. Eby, and E.L. Dolan. 2017. "Race and Gender Differences in Undergraduate Research Mentoring Structures and Research Outcomes." *CBE Life Sciences Education* 16(2):ar34. https://pubmed.ncbi.nlm.nih.gov/28550078.

AIR (American Institutes for Research). 2014. "The Role of Historically Black Colleges and Universities as Pathway Providers: Institutional Pathways to the STEM PhD Among Black Students." *Broadening Participation in STEM Graduate Education*. https://www.air.org/sites/default/files/downloads/report/Role%20of%20HBCUs%20in%20STEM%20PhDs%20for%20Black%20Students.pdf.

American Academy of Arts & Sciences. 2020. *The Perils of Complacency: America at a Tipping Point in Science & Engineering*. Cambridge, MA. https://www.amacad.org/publication/perils-of-complacency.

Baker, V.L., M.J. Pifer, and K.A. Griffin. 2014. "Mentor-Protégé Fit: Identifying and Developing Effective Mentorship across Identities in Doctoral Education." *International Journal for Researcher Development* 5(2):83-98. http://dx.doi.org/10.1108/IJRD-04-2014-0003.

Barbier, L., C. Wilson, and A. Kc. 2021. "SMD Summary Demographic Data." Presentation to the National Academies Committee on Increasing Diversity and Inclusion in the Leadership of Competed Missions. https://www.nationalacademies.org/event/08-12-2021/docs/DF47BBCCD1A00AB26A99C2D42FD76C00646CDF1F6033.

Batchelor, R.L., H. Ali, K.G. Gardner-Vandy, A.U. Gold, J.A. MacKinnon, and P.M. Asher. 2021. "Reimagining STEM Workforce Development as a Braided River." *Eos* 102. https://eos.org/opinions/reimagining-stem-workforce-development-as-a-braided-river.

Beasley, M.A. 2011. *Opting Out: Losing the Potential of America's Young Black Elite*. University of Chicago Press, Chicago, IL.

Bell, S.T., A.J. Villado, M.A. Lukasik, L. Belau, and A.L. Briggs. 2011. "Getting Specific About Demographic Diversity Variable and Team Performance Relationships: A Meta-Analysis." *Journal of Management* 37(3):709-743. https://doi.org/10.1177%2F0149206310365001.

Bernard, H.R. 2000. *Social Research Methods: Qualitative and Quantitative Approaches*, 2nd ed. Sage Publishing, Los Angeles, CA.

Bernard, R.E., and E.H.G. Cooperdock. 2018. "No Progress on Diversity in 40 Years." *Nature Geoscience* 11:292-295. https://www.nature.com/articles/s41561-018-0116-6.

Bertrand, M., and S. Mullainathan. 2004. "Are Emily and Greg More Employable Than Lakisha and Jamal? A Field Experiment on Labor Market Discrimination." *American Economic Review* 94(4):991-1013. https://www.aeaweb.org/articles?id=10.1257/0002828042002561.

Bielby, W.T. 2000. "Minimizing Workplace Gender and Racial Bias." *Contemporary Sociology* 29(1):120-129. https://doi.org/10.2307/2654937.

Blair-Loy, M., and E.A. Cech. 2022. *Misconceiving Merit: Paradoxes of Excellence and Devotion in Academic Science and Engineering.* University of Chicago Press, Chicago, IL.

Blyth, C.R. 1972. "On Simpson's Paradox and the Sure-Thing Principle." *Journal of the American Statistical Association* 67(338):364-366. https://doi.org/10.2307/2284382.

Bronfenbrenner, U. 1974. "Developmental Research, Public Policy, and the Ecology of Childhood." *Child Development* 45(1):1-5. https://doi.org/10.2307/1127743.

Bronfenbrenner, U. 1995. "Developmental Ecology Through Space and Time: A Future Perspective." Pp. 619-647 in *Examining Lives in Context: Perspectives on the Ecology of Human Development* (P. Moen, G.H. Elder, Jr., and K. Lüscher, eds.). American Psychological Association, Washington, DC. https://doi.org/10.1037/10176-018.

Byars-Winston, A.M., J. Branchaw, C. Pfund, P. Leverett, and J. Newton. 2015. "Culturally Diverse Undergraduate Researchers' Academic Outcomes and Perceptions of Their Research Mentoring Relationships." *International Journal of Science Education* 37(15):2533-2554. https://doi.org/10.1080/09500693.2015.1085133.

Cain, C.L., and E. Leahy. 2014. "Cultural Correlates of Gender Integration in Science." *Gender Work and Organization* 21(6):516-530. https://doi.org/10.1111/gwao.12052.

Calabrese Barton, A., H. Kang, E. Tan, T. O'Neill, and C. Brecklin. 2013. "Crafting a Future in Science: Tracing Middle School Girls' Identity Work Over Time and Space." *American Education Research Journal* 50(1):37-75. https://doi.org/10.3102%2F0002831212458142.

Campbell, L., S. Mehtani, M. Dozier, and J. Rinehart. 2013. "Gender-Heterogeneous Working Groups Produce Higher Quality Science." *PLOS ONE* 8(10):e79147. https://doi.org/10.1371/journal.pone.0079147.

Cannady, M.A., E. Greenwald, and K.N. Harris. 2014. "Problematizing the STEM Pipeline Metaphor: Is the STEM Pipeline Metaphor Serving Our Students and the STEM Workforce? *Science Education* 98(3):443-460. https://doi.org/10.1002/sce.21108.

Carlone, H.B. 2004. "The Cultural Production of Science in Reform-Based Physics: Girls' Access, Participation, and Resistance." *Journal of Research in Science Teaching* 41(4):392-414. https://doi.org/10.1002/tea.20006.

Cech, E.A., and M. Blair-Loy. 2019. "The Changing Career Trajectories of New Parents in STEM." *Proceedings of the National Academy of Sciences* 116(10):4182-4187. https://doi.org/10.1073/pnas.1810862116.

Cech, E.A., and L.T. O'Connor. 2018. "Like Second-Hand Smoke: The Toxic Effect of Workplace Flexibility Bias for Workers' Health." *Community, Work & Family* 20(5):543-572. https://doi.org/10.1080/13668803.2017.1371673.

Cech, E.A., and T.J. Waidzunas. 2021. "Systemic Inequalities for LGBTQ Professionals in STEM." *Science Advances* 7(3):eabe0933. https://www.science.org/doi/10.1126/sciadv.abe0933.

Centrella, J., M. New, and M. Thompson. 2019. "Leadership and Participation in NASA's Explorer-Class Missions." White paper submitted to *Astro2020: Decadal Survey on Astronomy and Astrophysics*. https://arxiv.org/abs/1909.10314.

CEOSE (Committee on Equal Opportunities in Science and Engineering). 2013. *2011-2012 Biennial Report to Congress: Broadening Participation in America's STEM Workforce*. National Science Foundation, Alexandria, VA. https://www.nsf.gov/od/oia/activities/ceose/reports/2011-2012-ceose-biennial-report-508.pdf.

CEOSE. 2017. *2015-2016 Biennial Report to Congress: Broadening Participation in America's STEM Workforce*. National Science Foundation, Alexandria, VA. https://www.nsf.gov/od/oia/activities/ceose/reports/2015-2016-ceose-biennial-report-508.pdf.

CEOSE. 2019. *2017-2018 Biennial Report to Congress: Investing in Diverse Community Voices*. National Science Foundation, Alexandria, VA. https://www.nsf.gov/od/oia/activities/ceose/reports/2017-2018-ceose-biennial-report-508.pdf.

Chakraverty, D. 2019. "Impostor Phenomenon in STEM: Occurrence, Attribution, and Identity." *Studies in Graduate and Postdoctoral Education* 10(1):2-20. https://doi.org/10.1108/SGPE-D-18-00014.

Cheng, H., G. Potvin, R. Khatri, L.H. Kramer, R.M. Lock, and Z. Hazari. 2018. "Examining Physics Identity Development Through Two High School Interventions." Paper presented at the Physics Education Research Conference, Washington, DC. http://dx.doi.org/10.1119/perc.2018.pr.Cheng.

Clance, P.R. 1985. *The Impostor Phenomenon: Overcoming The Fear That Haunts Your Success*. Peachtree Publishers, Atlanta, GA.

Cohen, J.J., B.A. Gabriel, and C. Terrell. 2002. "The Case for Diversity in the Health Care Workforce." *Health Affairs* 21(5):90-102. https://doi.org/10.1377/hlthaff.21.5.90.

Cole, E.R. 2021. "Intersectionality Framework and Its Importance in DEIA Work." Presentation to the National Academies Committee on Increasing Diversity and Inclusion in the Leadership of Competed Space Missions. https://www.nationalacademies.org/event/02-08-2021/docs/D3B1990EE01654A1446998C1A6A744682768622D21D0.

Cooper, B. 2015. "Intersectionality." Pp. 385-406 in *Oxford Handbook of Feminist Theory* (L. Disch and M. Hawkesworth, eds.). Oxford University Press, New York.

Correll, S.J. 2004. "Constraints into Preferences: Gender, Status, and Emerging Career Aspirations." *American Sociological Review* 69(1):93-113. https://doi.org/10.1177%2F000312240406900106.

Cosgrove, B. 2019. "NSF Policies on Harassment and Discrimination in NSF-Funded Awardee Programs." High Energy Physics Advisory Panel Presentation, National Science Foundation. https://science.osti.gov/-/media/hep/hepap/pdf/201905/Cosgrove-NSF_Diversity_Program_HEPAP201905.pdf.

Denizen, N.K., and Y.S. Lincoln. 2005. "Introduction: The Discipline and Practice of Qualitative Research." In *The SAGE Handbook of Qualitative Research* (N.K. Denizen and Y.S. Lincoln, eds.). Sage Publications, Los Angeles, CA.

Diaz-Garcia, C., A. Gonzalez-Moreno, and F.J. Saez-Martinez. 2013. "Gender Diversity Within R&D Teams: Its Impact on Radicalness of Innovation." *Innovation* 15(2):149-160. https://doi.org/10.5172/impp.2013.15.2.149.

Diverse. 2021. "Top 100 Degree Producers." *Diverse: Issues in Higher Education* 38(13). https://top100.diverseeducation.com.

DoD (Department of Defense). 2020. "DoD Science, Technology, Engineering, and Mathematics (STEM) Education Programs and Activities" DoDI 1025.11. Washington, DC. https://irp.fas.org/doddir/dod/i1025_11.pdf.

Douglass, J.A. 2000. "A Certain Future: Sputnik, American Higher Education, and the Survival of a Nation." Pp. 327-363 in *Reconsidering Sputnik: Forty Years Since the Soviet Satellite* (R.D. Launius, J.M. Logsdon, and R.W. Smith, eds.). Routledge, New York.

Duhigg, C. 2016. "What Google Learned from Its Quest to Build the Perfect Team." *The New York Times Magazine*, February 25. https://www.nytimes.com/2016/02/28/magazine/what-google-learned-from-its-quest-to-build-the-perfect-team.html.

Eaton, A.A., J.F. Saunders, R.K. Jacobson, and K .West. 2019. "How Gender and Race Stereotypes Impact the Advancement of Scholars in STEM: Professors' Biased Evaluations of Physics and Biology Post-Doctoral Candidates." *Sex Roles* 82:127-141. https://doi.org/10.1007/s11199-019-01052-w.

Ecklund, E.H., A.E. Lincoln, and T. Cassandra. 2012. "Gender Segregation in Elite Academic Science." Gender & Society 26(5):693-717. https://doi.org/10.1177%2F0891243212451904.

Edmondson, A. 1999. Psychological Safety and Learning Behavior in Work Teams." *Administrative Science Quarterly* 44(2):350-383. https://doi.org/10.2307%2F2666999.

Edmondson, A.C. 2004. "Psychological Safety, Trust, and Learning in Organizations: A Group-Level Lens." Pp. 239-272 in *Trust and Distrust in Organizations: Dilemmas and Approaches* (R.M. Kramer and K.S. Cook, eds.). Russell Sage Foundation, New York.

EOP (Executive Office of the President). 2019. "Executive Order 13872: Economic Empowerment of Asian Americans and Pacific Islanders" 84 FR 22321. *Federal Register* 84(95):22321-22325. https://www.federalregister.gov/d/2019-10398.

EOP. 2016. "Presidential Memorandum—Promoting Diversity and Inclusion in the National Security Workforce." https://obamawhitehouse.archives.gov/the-press-office/2016/10/05/presidential-memorandum-promoting-diversity-and-inclusion-national.

EOP. 2021. *Best Practices for Diversity and Inclusion in STEM Education and Research: A Guide for and by Federal Agencies*. A report by the Interagency Working Group on Inclusion in STEM, the Federal Coordination in STEM Education Subcommittee, and the Committee on STEM Education of the National Science and Technology Council. Washington, DC. https://www.whitehouse.gov/wp-content/uploads/2021/09/091621-Best-Practices-for-Diversity-Inclusion-in-STEM.pdf.

EOP and OPM (Office of Personnel Management). 2016. *Reducing the Impact of Bias in the STEM Workforce: Strengthening Excellence and Innovation*. A report of the Interagency Policy Group on Increasing Diversity in the STEM Workforce by Reducing the Impact of Bias. Washington, DC. https://obamawhitehouse.archives.gov/sites/default/files/microsites/ostp/ostp-opm_bias_mitigation_report__20161129.pdf.

Erosheva, E.A., S. Grant, M.C. Chen, M.D. Lindner, R.K. Nakamura, and C.J. Lee. 2020. "NIH Peer Review: Criterion Scores Completely Account for Racial Disparities in Overall Impact Scores." *Science Advances* 6(23):eaaz4868. https://www.science.org/doi/abs/10.1126/sciadv.aaz4868.

Estrada, M., M. Burnett, A.G. Campbell, P.B. Campbell, W.F Denetclaw, C.G. Gutiérrez, S. Hurtado, et al. 2016. "Improving Underrepresented Minority Student Persistence in STEM." *CBE-Life Sciences Education* 15(3):1-10. https://doi.org/10.1187/cbe.16-01-0038.

Estrada, M., A. Woodcock, P.R. Hernandez, and P.W. Schultz. 2011. "Toward a Model of Social Influence That Explains Minority Student Integration into the Scientific Community." *Journal of Educational Psychology* 103(1):206-222. https://psycnet.apa.org/doi/10.1037/a0020743.

FCA (Family Caregiver Alliance). 2012. Selected Caregiver Statistics (Fact Sheet). https://www.caregiver.org/selected-caregiver-statistics.

Feldman, M.P., L. Lanahan, and I.V. Lendel. 2014. "Experiments in the Laboratories of Democracy: State Scientific Capacity Building." *Economic Development Quarterly* 28(2):107-131. https://doi.org/10.1177%2F0891242413490018.

Fiore, S.M., D.R. Carter, and R. Asencio. 2015. "Conflict, Trust, and Cohesion: Examining Affective and Attitudinal Factors in Science Teams." Pp. 271-301 in *Team Cohesion: Advances in Psychological Theory, Methods and Practice* (E. Salas, W.B. Vessey, and A.X. Estrada, eds.). Emerald Group Publishing, Bingley, United Kingdom.

Fisher, A.J., R. Mendoza-Denton, C. Patt, I. Young, A. Eppig, R.L. Garrell, D.C. Rees, T.W. Nelson, and M.A. Richards. 2019. "Structure and Belonging: Pathways to Success for Underrepresented Minority and Women PhD Students in STEM Fields." *PLOS ONE* 14(1):e0209279. https://doi.org/10.1371/journal.pone.0209279.

Fiske, S.T., A.J.D. Cuddy, P. Glick, and J. Xu. 2002. "A Model of (Often Mixed) Stereotype Content: Competence and Warmth Respectively Follow from Perceived Status and Competition." *Journal of Personality and Social Psychology* 82(6):878-902. https://doi.org/10.1037/0022-3514.82.6.878.

Freeman, R.B., and W. Huang. 2014a. "Collaboration: Strength in Diversity." *Nature* 513(305). https://www.nature.com/articles/513305a.

Freeman, R.B., and W. Huang. 2014b. "Collaborating with People Like Me: Ethnic Co-Authorship Within the US." NBER Working Paper Series 19905. National Bureau of Economic Research, Cambridge, MA. https://www.nber.org/papers/w19905.

Gammie, A. 2021. "Overview of NIGMS Diversity Programs." Presentation to the National Academies Committee on Increasing Diversity and Inclusion in the Leadership of Competed Space Missions on April 27. https://www.nationalacademies.org/event/04-27-2021/docs/DD5FBE84CDB8E77306EA881BDA579C7E17ABEFEBA5B4.

Gasman, M., B. Baez, N. Drezner, K. Sedgwick, and C. Tudico. 2007. "Historically Black Colleges and Universities: Recent Trends." American Association of University Professors. Washington, DC. https://www.aaup.org/report/historically-black-colleges-and-universities-recent-trends.

Gazley, J.L., R. Remich, M.E. Naffziger-Hirsch, J. Keller, P.B. Campbell, and R. McGee. 2014. "Beyond Preparation: Identity, Cultural Capital, and Readiness for Graduate School in the Biomedical Sciences." *Journal of Research in Science Teaching* 51:1021-1048. https://doi.org/10.1002/tea.21164.

Ginther, D.K., W.T. Schaffer, J. Schnell, B. Masimore, F. Liu, L.L. Haak, and R. Kington. 2011. "Race, Ethnicity, and NIH Research Awards." *Science* 333(6045):1015-1019. https://doi.org/10.1126/science.1196783.

Goldin, C., and C. Rouse. 2000. "Orchestrating Impartiality: The Impact of 'Blind' Auditions on Female Musicians." *American Economic Review* 90(4):715-741. https://www.aeaweb.org/articles?id=10.1257/aer.90.4.715.

Goulden, M., M.A. Mason, and K. Frasch. 2011. "Keeping Women in the Science Pipeline." *The ANNALS of the American Academy of Political and Social Science* 638(1):141-162. https://doi.org/10.1177%2F0002716211416925.

Graham, M.J., J. Frederick, A. Byars-Winston, A. Hunter, and J. Handelsman. 2013. "Increasing Persistence of College Students in STEM." *Science* 341(6153):1455-1456. https://www.science.org/doi/10.1126/science.1240487.

Griffin, K.A. 2018. "Addressing STEM Culture and Climate to Increase Diversity in STEM Disciplines." *Higher Education Today.* Blog. American Council on Education. https://www.higheredtoday.org/2018/04/23/addressing-stem-culture-climate-increase-diversity-stem-disciplines.

Guba, E.G., and Y.S. Lincoln. 1994. "Competing Paradigms in Qualitative Research." Pp. 105-117 in *Handbook of Qualitative Research* (N.K. Denizen and Y.S. Lincoln, eds.). Sage Publishing, Los Angeles, CA.

Hall, K.L., D. Stokols, R.P. Moser, B.K. Taylor, M.D. Thornquist, L.C. Nebeling, C.C. Ehret, et al. 2008. "The Collaboration Readiness of Transdisciplinary Research Teams and Centers: Findings from the National Cancer Institute's TREC Year-One Evaluation Study." *American Journal of Preventive Medicine* 35(2):S161-S172. https://doi.org/10.1016/j.amepre.2008.03.035.

Hall, K.L., A.L. Vogel, G.C. Huang, K.J. Serrano, E.L. Rice, S.P. Tsakraklides, and S.M. Fiore. 2018. "The Science of Team Science: A Review of the Empirical Evidence and Research Gaps on Collaboration in Science." *American Psychologist* 73(4):532-548. https://psycnet.apa.org/doi/10.1037/amp0000319.

Hamden, E., N. Cabrera-Salazar, E. Gertsen, and M. New. 2019. "Launchpad." PI Launchpad, Tucson, AZ. https://science.nasa.gov/science-pink/s3fs-public/atoms/files/PI%20Workbook%20-%20FINAL.pdf.

Helba, C., A.M. Porter, S. Nicholson, and R. Ivie. 2019. "Women Among Physics and Astronomy Faculty: Results from the 2018 Academic Workforce Survey." AIP Focus On Report. American Institute of Physics Statistical Research Center, College Park, MD. https://www.aip.org/statistics/reports/women-among-physics-and-astronomy-faculty.

Hofstra, B., V.V. Kulkarni, S.M. Galvez, B. He, D. Jurafsky, and D.A. McFarland. 2020. "The Diversity-Innovation Paradox in Science." *Proceedings of the National Academy of Sciences* 117(17):9284-9291. https://doi.org/10.1073/pnas.1915378117.

Holgersson, C., and L. Romani. 2020. "Tokenism Revisited: When Organizational Culture Challenges Masculine Norms, the Experience of Token Is Transformed." *European Management Review* 17:649-661. https://doi.org/10.1111/emre.12385.

Hong, S., and S.E. Page. 2004. "Groups of Diverse Problem Solvers Can Outperform Groups of High-Ability Problem Solvers." *Proceedings of the National Academy of Sciences* 101(46):16385-16389. https://doi.org/10.1073/pnas.0403723101.

Hoppe, T.A., A. Litovitz, K.A. Willis, R.A. Meseroll, M.J. Perkins, B.I. Hutchins, A.F. Davis, M.S. Lauer, H.A. Valantine, J.M. Anderson, and G.M. Santangelo. 2019. "Topic Choice Contributes to the Lower Rate of NIH Awards to African-American/Black Scientists." *ScienceAdvances* 5(10). https://www.science.org/doi/10.1126/sciadv.aaw7238.

Hubbard, S.M., and F.K. Stage. 2010. "Identifying Comprehensive Public Institutions That Develop Minority Scientists." *New Directions for Institutional Research* 2010(148):53-62. https://doi.org/10.1002/ir.361.

Hurtado, S., J.F. Milem, A. Clayton-Pedersen, and W.R. Allen. 1998. "Enhancing Campus Climates for Racial/Ethnic Diversity: Educational Policy and Practice." *The Review of Higher Education* 21(3):279-302. https://doi.org/10.1353/rhe.1998.0003.

Ivie, R., G. Anderson, and S. White. 2014. *African Americans & Hispanics Among Physics and Astronomy Faculty: Results from the 2012 Survey of Physics & Astronomy Degree-Granting Departments*. AIP Focus On Report February 2010. American Institute of Physics Statistical Research Center, College Park, MD. https://www.aip.org/statistics/reports/african-americans-hispanics-among-physics-astronomy-faculty-0.

Ivie, R., and F. Bagenal. 2011. "Survey of the Planetary Science Workforce." American Institute of Physics Statistical Research Center, College Park, MD. https://lasp.colorado.edu/home/mop/files/2015/08/PlanetSciSurvey.pdf.

Ivie, R., A. Ephraim, and S. White. 2010. *African Americans & Hispanics Among Physics and Astronomy Faculty: Results from the 2008 Survey of Physics & Astronomy Degree-Granting Departments*. AIP Focus On Report February 2010. American Institute of Physics Statistical Research Center, College Park, MD. https://www.aip.org/statistics/reports/african-americans-hispanics-among-physics-astronomy-faculty.

Ivie, R., and K. Nies. 2005. *Women in Physics and Astronomy, 2005* AIP Report R-430.02. American Institute of Physics Statistical Research Center, College Park, MD. https://www.aip.org/sites/default/files/statistics/women/women-pa-05.pdf.

Ivie, R., and S. White. 2020. *Measuring and Analyzing the Gender Gap in Science Through the Global Survey of Scientists*. AIP Focus On Report January 2020. American Institute of Physics Statistical Research Center, College Park, MD. https://www.aip.org/sites/default/files/2022-02/Measuregendrgap.pdf.

Jensen, H., and L. Pappas. 2021. "Explore Solar System & Beyond: NASA Announcement of Opportunity Science Team Demographics." Presentation to the Committee on Increasing Diversity and Inclusion in the Leadership of Competed Space Missions. National Aeronautics and Space Administration, Washington, DC. https://www.nationalacademies.org/event/06-162021/docs/DE434EB902FF6358FD08B572058EF5F4D85D88E56979.

Jones, S. 2021. "Mathematical and Physical Sciences at the National Science Foundation." Presentation to the National Academies Committee on Increasing Diversity and Inclusion in the Leadership of Competed Space Missions on April 27.

Joshi, A. 2014. "By Whom and When Is Women's Expertise Recognized? The Interactive Effects of Gender and Education in Science and Engineering Teams." *Administrative Science Quarterly* 59(2):202-239. https://doi.org/10.1177%2F0001839214528331.

Kachchaf, R., L. Ko, A. Hodari, and M. Ong. 2015. "Career-Life Balance for Women of Color: Experiences in Science and Engineering Academia." *Journal of Diversity in Higher Education* 8(3):175-191. https://psycnet.apa.org/doi/10.1037/a0039068.

Kafafi, Z.H. "NSF 08-062." National Science Foundation, Arlington, VA. https://www.nsf.gov/pubs/2008/nsf08062/nsf08062.pdf.

Kanter, R.M. 1977a. *Men and Women of the Corporation*. Basic Books, New York.

Kanter, R.M. 1977b. "Some Effects of Proportions on Group Life." Pp. 53-78 in *The Gender Gap in Psychotherapy*. Springer, Boston, MA.

Khanna, G. 2021. "How Higher Education Became an Important US Export." *Issues in Science and Technology* 38(1):30-33. National Academies of Sciences, Engineering, and Medicine and Arizona State University. https://issues.org/international-students-us-higher-education-statistics.

Koch, L. 2021. "Increasing Diversity, Equity, Inclusion, and Accessibility at NOAA." Presentation to the National Academies Committee on Increasing Diversity and Inclusion in the Leadership of Competed Space Missions on June 8.

Kuh, G.D. 2008. *High Impact Educational Practices: What They Are, Who Has Access to Them, and Why They Matter*. American Association for Colleges & Universities, Washington, DC.

Lamont, M., and A. Swidler. 2014. "Methodological Pluralism and the Possibilities and Limits of Interviewing." *Qualitative Sociology* 37(2):153-171. https://doi.org/10.1007/s11133-014-9274-z.

Landivar, L.C. 2013. *The Relationship Between Science and Engineering Education and Employment in STEM Occupations*. American Community Survey Reports, ACS-23. U.S. Census Bureau, Washington, DC. https://www2.census.gov/library/publications/2013/acs/acs-23.pdf.

Larose, S., C.F. Ratelle, F. Guay, C. Senécal, and M. Harvey. 2006. "Trajectories of Science Self-Efficacy Beliefs During the College Transition and Academic and Vocational Adjustment in Science and Technology Programs." *Educational Research and Evaluation* 12(4):373-393. https://doi.org/10.1080/13803610600765836.

Lebaron, M., and V. Pillay. 2006. *Conflict Across Cultures: A Unique Experience of Bridging Differences*. Intercultural Press, Boston, MA.

Lee, C.J., S. Grant, and E.A. Erosheva. 2020. "Alternative Grant Models Might Perpetuate Black-White Funding Gaps." *Lancet* 396(10256):955-956. https://doi.org/10.1016/S0140-6736(20)32018-3.

Leslie, S.-J., A. Cimpian, M. Meyer, and E. Freeland. 2015. "Expectations of Brilliance Underlie Gender Distributions Across Academic Disciplines." *Science* 347(6219):262-265. https://doi.org/10.1126/science.1261375.

Lewis, K.L., J.G. Stout, N.D. Finkelstein, S.J. Pollock, A. Miyake, G.L. Cohen, and T.A. Ito. 2017. "Fitting in to Move Forward: Belonging, Gender, and Persistence in the Physical Sciences, Technology, Engineering, and Mathematics (pSTEM)." *Psychology of Women Quarterly* 41(4):420-436. https://doi.org/10.1177/0361684317720186.

Lipsy, S. 2021. "PI-Led Mission Proposal Process." Presentation to the National Academies Committee on Increasing Diversity and Inclusion in the Leadership of Competed Missions on February 8.

Lonsdale, C., F.R. Schwab, and G. Hunt. 2016. *Gender-Related Systematics in NRAO and ALMA Proposal Review Processes*. National Radio Astronomy Observatory, Charlottesville, VA. https://science.nrao.edu/science/reports/gender-systematics-in/gender-related-systematics-in-the-nrao-and-alma-proposal-review-processes.

López, C.M., C. Margherio, L.M. Abraham-Hilaire, and C. Feghali-Bostwick. 2018. "Gender Disparities in Faculty Rank: Factors That Affect Advancement of Women Scientists at Academic Medical Centers." *Social Sciences* 7(4):62. https://doi.org/10.3390/socsci7040062.

Majer, J.M. 2009. "Self-Efficacy and Academic Success Among Ethnically Diverse First-Generation Community College Students." *Journal of Diversity in Higher Education* 2(4):243-250. https://doi.org/10.1037/a0017852.

Malcom-Piqueux, L. 2020. "Transformation in the U.S. Higher Education System: Implications for Racial Equity. Paper commissioned by the National Academies for the Symposium on Imagining the Future of Undergraduate STEM Education." https://www.nationalacademies.org/event/10-21-2020/docs/DC5839F575628FD0BF22011A332693AFB8D99D306381.

McCrary, V. 2021. "NSB Vision 2030: Priorities for STEM: A Vision for American's S&E Enterprise." Presentation to the National Academies Committee on Increasing Diversity and Inclusion in the Leadership of Competed Space Missions, National Science Board, Alexandria, VA. https://www.nationalacademies.org/event/06-08-2021/docs/DFEF7DD190113474FC06B3DAB1C9BF0EE35116AE483E.

McIlwee, J.S., and J.G. Robinson. 1992. *Women in Engineering: Gender, Power, and Workplace Culture*. State University of New York Press, Albany.

Mehta, C.R., and N.R. Patel. 1983. "A Network Algorithm for Performing Fisher's Exact Test in r× c Contingency Tables." *Journal of the American Statistical Association* 78(382):427-434. https://doi.org/10.2307/2288652.

Melkers, J., and Y. Wu. 2009. "Evaluating the Improved Research Capacity of EPSCoR States: R&D Funding and Collaborative Networks in the NSF EPSCoR Program." *Review of Policy Research* 26(6):761-782. https://doi.org/10.1111/j.1541-1338.2009.00415.x.

Mijid, N. 2014. "Why Are Female Small Business Owners in the United States Less Likely to Apply for Bank Loans Than Their Male Counterparts?" *Journal of Small Business & Entrepreneurship* 27(2):229-249. https://doi.org/10.1080/08276331.2015.1012937.

Moss-Racusin, C.A., J.F. Dovidio, V.L. Brescoll, M.J. Graham, and J. Handelsman. 2012. "Science Faculty's Subtle Gender Biases Favor Male Students." *Proceedings of the National Academy of Sciences* 109(41):16474-16479. https://doi.org/10.1073/pnas.1211286109.

Moss-Racusin, C.A., and L.A. Rudman. 2010. "Disruptions in Women's Self-Promotion: The Backlash Avoidance Model." *Psychology of Women Quarterly* 34(2):186-202. https://doi.org/10.1111%2Fj.1471-6402.2010.01561.x.

Mulvey, P.J., S. Nicholson, and J. Pold. 2021. *Trends in Physics PhDs: Results from the 2019 Survey of Enrollments and Degrees and the Degree Recipient Follow-Up Survey for the Classes of 2017 and 2018*. AIP Focus On Report February 2021. American Institute of Physics Statistical Research Center, College Park, MD. https://www.aip.org/statistics/reports/trends-physics-phds-171819.

Nakamura, R., L.S. Mann, M.D. Lindner, J. Braithwaite, M.C. Chen, A. Vancea, N. Byrnes, V. Durrant, and B. Reed. 2021. "An Experimental Test of the Effects of Redacting Grant Applicant Identifiers on Peer Review Outcomes." *eLife* 10:e71368. https://elifesciences.org/articles/71368.

NAS, NAE, and IOM (National Academy of Sciences, National Academy of Engineering, and Institute of Medicine). 2007. *Beyond Bias and Barriers: Fulfilling the Potential of Women in Academic Science and Engineering*. The National Academies Press, Washington, DC. https://doi.org/10.17226/11741.

NAS, NAE, and IOM. 2011. *Expanding Underrepresented Minority Participation: America's Science and Technology Talent at the Crossroads*. The National Academies Press, Washington, DC. https://doi.org/10.17226/12984.

NASA (National Aeronautics and Space Administration). 2007. "SMEX 2007 Announcement of Opportunity." https://explorers.larc.nasa.gov/CommAnnouncementFAQ.html.

NASA. 2008. National Aeronautics and Space Act of 1958, as Amended. Washington, DC. https://history.nasa.gov/spaceact-legishistory.pdf.

NASA. 2010. "Standard PI-Led Mission Template: Guidelines and Criteria for the Phase A Concept Study." Science Mission Directorate. https://soma.larc.nasa.gov/StandardAO/sao_templates.html.

NASA. 2012. "Technology Readiness Levels." https://www.nasa.gov/directorates/heo/scan/engineering/technology/technology_readiness_level.

NASA. 2015a. "Criteria and Factors." Section 1872.402 in *NASA Federal Acquisitions Regulation Supplement*. Washington, DC. https://www.hq.nasa.gov/office/procurement/regs/NFS.pdf.

NASA. 2015b. *Promising Practices for Equal Opportunity, Diversity, and Equity*. Office of Diversity and Equal Opportunity. Washington, DC. https://odeo.hq.nasa.gov/documents/PromPract_8-20-15_TAGGED.pdf.

NASA. 2016. *NASA 2016 Diversity and Inclusion Strategic Implementation Plan*. Washington, DC. https://odeo.hq.nasa.gov/documents/FINAL_DI_Plan_8-15-16_TAGGED.pdf.

NASA. 2018. "Standard AO information." https://soma.larc.nasa.gov/StandardAO/sao_templates.html.

NASA. 2019. "NASA Program/Project Life Cycle." Section 3.0 in *Systems Engineering Handbook*. Washington, DC. https://www.nasa.gov/seh/3-project-life-cycle.

NASA. 2020a. "Earth Venture Mission-3 (EVM-3) Earth System Science Pathfinder Program." NNH21ZDA002O. NASA Solicitation and Proposal Integrated Review and Evaluation System. https://nspires.nasaprs.com.

NASA. 2020b. *Science 2020-2024: A Vision for Scientific Excellence*. NASA Headquarters, Washington, DC. https://science.nasa.gov/science-red/s3fs-public/atoms/files/2020-2024_Science.pdf.

NASA. 2021a. "2019 Heliophysics Explorers Program (HP) Medium-Class Explorer (MIDEX) Concept Study Report (CSR) Evaluation Plan." Announcement of Opportunity NNH19ZDA013O. https://explorers.larc.nasa.gov/HPMIDEX/pdf_files/HPMIDEX19_DS_Eval_Plan_20210804.pdf.

NASA. 2021b. *Guidebook for Proposers Responding to a NASA Notice of Funding Opportunity (NOFO)*, Revised as of April 28, 2021. Washington, DC._https://www.nasa.gov/sites/default/files/atoms/files/may_2021_ed._nasa_guidebook_for_proposers.pdf.

NASA. 2021d. *NASA Space Flight Program and Project Management Requirements*. NPR 7120.5F. Washington, DC. https://nodis3.gsfc.nasa.gov/displayDIR.cfm?t=NPR&c=7120&s=5F.

NASA. 2021e. "National Aeronautics and Space Administration Policy Statement on Diversity, Equity, Inclusion, and Accessibility for NASA's Workforce and Workplaces." https://www.nasa.gov/sites/default/files/atoms/files/2021_administrator_policy_statement_on_deia_at_nasa_9-28-21_tagged.pdf.

NASA. 2021f. "Request for Information on Advancing Racial Equity and Support for Underserved Communities in NASA Programs, Contracts and Grants Process." *Federal Register* 86(113):31735-31738. https://www.govinfo.gov/content/pkg/FR-2021-06-15/pdf/2021-12668.pdf.

NASA. 2021g. "Request for Information: Adding Inclusion, Diversity, Equity, and Accessibility (IDEA) Requirements to NASA Announcements of Opportunity." NNH21ZDA019L. https://nspires.nasaprs.com.

NASA. 2021h. "Request for Information: Proposed New Language on Inclusion, Diversity, Equity, and Accessibility for Science Mission Directorate Announcements of Opportunity" NNH21XDA019L. http://nspires.nasaprs.com.

NASA. 2021i. "Writing Successful Mission Proposals." New Principal Investigator (PI) Resources. https://science.nasa.gov/researchers/new-pi-resources.

NASEM (National Academies of Sciences, Engineering, and Medicine). 2016. *Barriers and Opportunities for 2-Year and 4-Year STEM Degrees: Systemic Change to Support Students' Diverse Pathways*. The National Academies Press, Washington, DC. https://doi.org/10.17226/21739.

NASEM. 2017a. *Powering Science—NASA's Large Strategic Science Missions*. The National Academies Press, Washington, DC. https://doi.org/10.17226/24857.

NASEM. 2017b. *Undergraduate Research Experiences for STEM Students: Successes, Challenges, and Opportunities*. The National Academies Press, Washington, DC. https://doi.org/10.17226/24622.

NASEM. 2019a. *Minority Serving Institutions: America's Underutilized Resource for Strengthening the STEM Workforce*. The National Academies Press, Washington, DC. https://doi.org/10.17226/25257.

NASEM. 2019b. *The Science of Effective Mentorship in STEMM*. The National Academies Press, Washington, DC. https://doi.org/10.17226/25568.

NASEM. 2020a. *The Impacts of Racism and Bias on Black People Pursuing Careers in Science, Engineering, and Medicine: Proceedings of a Workshop*. The National Academies Press, Washington, DC. https://doi.org/10.17226/25849.

NASEM. 2020b. *NASA's Science Activation Program: Achievements and Opportunities*. The National Academies Press, Washington, DC. https://doi.org/10.17226/25569.

NASEM. 2021. *Pathways to Discovery in Astronomy and Astrophysics for the 2020s*. The National Academies Press, Washington, DC. https://doi.org/10.17226/26141.

NCES (National Center for Education Statistics). 2021. "Finance Section of IPEDS Survey Components." In *Integrated Post-secondary Education Data System (IPEDS)*. https://nces.ed.gov/ipeds/use-the-data.

Newman, C.B., and M.B. Jackson. 2013. "Collaborative Partnerships in Engineering Between Historically Black Colleges and Universities and Predominantly White Institutions." Pp. 181-191 in *Fostering Success of Ethnic and Racial Minorities in STEM: The Role of Minority Serving Institutions* (R.T. Palmer, D.C. Maramba, and M. Gasman, eds.). Routledge, New York.

Nielsen, M.W., S. Alegria, L. Börjeson, H. Etzkowitz, H.J. Falk-Krzesinski, A. Joshi, and L. Schiebinger. 2017. "Opinion: Gender Diversity Leads to Better Science." *Proceedings of the National Academy of Sciences* 114(8):1740-1742. https://doi.org/10.1073/pnas.1700616114.

NIH (National Institutes of Health). 2019a. "NIH Workplace Climate and Harassment Survey." https://diversity.nih.gov/building-evidence/harassment-survey.

NIH. 2019b. "Supporting a Safe and Respectful Workplace: NIH Actions and Oversight to Address Harassment." https://grants.nih.gov/grants/policy/harassment/actions-oversight.htm.

NIH. 2020. "NIH Process for Handling Allegations of Harassment on an NIH-Funded Project at a Recipient Institution." https://grants.nih.gov/grants/policy/harassment/actions-oversight/allegation-process.htm.

NIH. 2021a. "NIH Grants Policy Statement." Revised. Office of Extramural Research. https://grants.nih.gov/grants/policy/nihgps/HTML5/section_4/4.1.2_civil_rights_protections.htm#Educatio.

NIH. 2021b. "NIH Workforce Profile." https://www.edi.nih.gov/people/resources/advancing-racial-equity/nih-workforce-profile-fy21q02.

NRC (National Research Council). 2006. *Principal-Investigator-Led Missions in the Space Sciences*. The National Academies Press, Washington, DC. https://doi.org/10.17226/11530.

NRC. 2014. *Solar and Space Physics: A Science for a Technological Society: An Overview*. The National Academies Press, Washington, DC. https://doi.org/10.17226/18974.

NRC. 2015. *Enhancing the Effectiveness of Team Science*. The National Academies Press, Washington, DC. https://doi.org/10.17226/19007.

NSB (National Science Board). 2022. "The State of US Science and Engineering 2022." https://ncses.nsf.gov/pubs/nsb20221.

NSF (National Science Foundation). 1997. "NSF to Adopt New Merit Review Criteria." https://www.nsf.gov/news/news_summ.jsp?cntn_id=102789.

NSF. 2011. "Balancing the Scale: NSF's Career-Life Balance Initiative." https://www.nsf.gov/career-life-balance/brochure.pdf.

NSF. 2018. *Doctorate Recipients from U.S. Universities: 2016*. Report NSF 18-304. National Center for Science and Engineering Statistics, Alexandria, VA. https://www.nsf.gov/statistics/2018/nsf18304.

NSF. 2019. *Doctorate Recipients from U.S. Universities: 2019*. Report NSF 21-308. National Center for Science and Engineering Statistics. Alexandria, VA. https://ncses.nsf.gov/pubs/nsf21308.

NSF. 2020. *The National Science Foundation's Merit Review Process: Fiscal Year 2019 Digest*. Alexandria, VA. https://www.nsf.gov/nsb/publications/2020/merit_review/FY-2019/nsb202038.pdf.

NSF. 2021a. "FAQs Regarding NSF's Award Term and Condition Entitled, *Notification Requirements Regarding Sexual Harassment, Other Forms of Harassment, or Sexual Assault*." https://www.nsf.gov/od/oecr/docs/Sexual_Harassment_FAQs.pdf.

NSF. 2021b. "Higher Education and Research Development Survey (HERD): FY2020." Data tables. NSF 22-311. National Center for Science and Engineering Statistics. https://ncses.nsf.gov/pubs/nsf22311.

NSF. 2021c. "Survey of Federal Funds for Research and Development." NSF 21-32. National Center for Science and Engineering Statistics. https://www.nsf.gov/statistics/srvyfedfunds.

NSF. 2021d. *Women, Minorities and Persons with Disabilities in Science and Engineering: 2021*. Special Report NSF 21-321. Alexandria, VA. https://ncses.nsf.gov/wmpd.

Ochoa, E., and V.R. McCrary. 2021. "Cultivating America's STEM Talent Must Begin at Home." *Issues in Science and Technology*, November 30. https://issues.org/cultivating-stem-talent-begins-at-home-ochoa-mccrary/#.Ynq0NcqeF7c.link.

Olson, M.A., and R.H. Fazio. 2003. "Relations Between Implicit Measures of Prejudice: What Are We Measuring?" *Psychological Science* 14(6):636-639. https://doi.org/10.1046/j.0956-7976.2003.psci_1477.x.

Ong, M., C. Wright, L. Espinosa, and G. Orfield. 2011. "A Synthesis of Empirical Research on Undergraduate and Graduate Women of Color in Science, Technology, Engineering, and Mathematics." *Harvard Educational Review* 81(2):172-209. https://doi.org/10.17763/haer.81.2.t022245n7x4752v2.

Page, S. 2007. *The Difference: How the Power of Diversity Creates Better Groups, Firms, Schools, and Societies*. Princeton University Press, Princeton, NJ.

Page, S.E. 2017. *The Diversity Bonus How Great Teams Pay Off in the Knowledge Economy*. Princeton University Press, Princeton, NJ.

Patat, F. 2016. "Gender Systematics in Telescope Time Allocation at ESO." *The Messenger* 165:2-9. https://www.eso.org/sci/publications/messenger/archive/no.165-sep16/messenger-no165-2-9.pdf.

Pinquart, M., and S. Sörensen. 2005. "Ethnic Differences in Stressors, Resources, and Psychological Outcomes of Family Caregiving: A Meta-Analysis." *The Gerontologist* 45:90-106. https://doi.org/10.1093/geront/45.1.90.

Pohlhaus, J.R., J. Jiang, R.M. Wagner, W.T. Schaffer, and V.W. Pinn. 2011. "Sex Differences in Application, Success, and Funding Rates for NIH Extramural Programs." *Academic Medicine* 86(6):759-767. https://doi.org/10.1097/ACM.0b013e31821836ff.

Pold, J., and R. Ivie. 2018. *Workforce Survey of 2018 US AAS Members: Summary Results*. American Institute of Physics Statistical Research Center, College Park, MD. https://aas.org/sites/default/files/2019-10/AAS-Members-Workforce-Survey-final.pdf.

Porter, A.M. 2019. *Physics PhDs Ten Years Later: Success Factors and Barriers in Career Paths: Results from the PhD Plus 10 Study*. AIP Focus On Report December 2019. American Institute of Physics Statistical Research Center, College Park, MD. https://www.aip.org/statistics/reports/physics-phds-ten-years-later-success-factors-and-barriers-career-paths.

Porter, AM., and R. Ivie. 2019. *Women in Physics andAstronomy, 2019*. Focus On Report March 2019. American Institute of Physics Statistical Research Center, College Park, MD. https://www.aip.org/statistics/reports/women-physics-and-astronomy-2019.

Porter, A.M., S. White, and J. Dollison. 2020. *2020 Survey of the Planetary Science Workforce*. American Institute of Physics Statistical Research Center, College Park, MD. https://dps.aas.org/sites/dps.aas.org/files/reports/2020/Results_from_the_2020_Survey_of_the_Planetary_Science_Workforce.pdf.

Posselt, J.R. 2020. *Equity in Science: Representation, Culture, and the Dynamic of Change in Graduate Education*. Stanford University Press, Palo Alto, CA. https://www.sup.org/books/title/?id=31068.

Rainey, K., M. Dancy, R. Mickelson, E. Stearns, and S. Moller. 2018. "Race and Gender Differences in How Sense of Belonging Influences Decisions to Major in STEM." *International Journal of STEM Education* 5(1):10. https://doi.org/10.1186/s40594-018-0115-6.

Rankin, S.R., and R.D. Reason. 2005. "Differing Perceptions: How Students of Color and White Students Perceive Campus Climate for Underrepresented Groups." *Journal of College Student Development* 46(1):43-61. https://doi.org/10.1353/csd.2005.0008.

Rathbun, J.A., S. Diniega, L.C. Quick, D.H. Grinspoon, S.M. Hörst, E.S. Lakdawalla, K.E. Mandt, et al. 2018. "The Planetary Science Workforce: Who Is Missing?" LPI Contrib. No. 2083 for the 49th Lunar and Planetary Science Conference 2018. https://www.hou.usra.edu/meetings/lpsc2018/pdf/2668.pdf.

Remich, R., M.E. Naffziger-Hirsch, J.L. Gazley, and R. McGee. 2016. "Scientific Growth and Identity Development During a Postbaccalaureate Program: Results from a Multisite Qualitative Study." *CBE Life Sciences Education* 15(3):ar25. https://doi.org/10.1187/cbe.16-01-0035.

Reskin, B.F. 2003. "Including Mechanisms in Our Models of Ascriptive Inequality: 2002 Presidential Address." *American Sociological Review* 68(1):1-21. https://doi.org/10.2307/3088900.

Risman, B.J. 2004. "Gender as a Social Structure: Theory Wrestling with Activism." *Gender & Society* 18(4):429-450. https://doi.org/10.1177/0891243204265349.

Rittmayer, A.D., and M.E. Beier. 2009. "Overview: Self-Efficacy in STEM." In *Applying Research to Practice (ARP) Resources* (B. Bogue and E. Cady, eds.). http://aweonline.org/arp_selfefficacy_overview_122208.pdf.

Rivera-Valentín, E.G., J. Rathbun, J.T. Keane, K.L. Lynch, C. Richey, S. Diniega, and J. Vertesi. 2020. "Who Is Missing in Planetary Science?: A Demographic Study of the Planetary Science Workforce." White paper submitted to the Planetary Science and Astrobiology Decadal Survey 2023-2032. *Bulletin of the AAS* 53(4). https://doi.org/10.3847/25c2cfeb.968ed505.

Rock, D., and H. Grant. 2016. "Why Diverse Teams Are Smarter." *Harvard Business Review* 4(4):2-5.

Roth, L.M. 2004. "The Social Psychology of Tokenism: Status and Homophily Processes on Wall Street." *Sociological Perspectives* 47(2):189-214. https://doi.org/10.1525%2Fsop.2004.47.2.189.

Rudman, L.A., and Glick, P. 2001. "Prescriptive Gender Stereotypes and Backlash Toward Agentic Women." *Journal of Social Issues* 57(4):743-762. https://doi.org/10.1111/0022-4537.00239.

Rudolph, A.L. 2019. "Cal-Bridge: Creating Pathways to the PhD for Underrepresented Students in Physics and Astronomy." *Physics Today* 72(10):50-57. https://physicstoday.scitation.org/doi/10.1063/PT.3.4319.

Sakimoto, P.J., and J.D. Rosenthal. 2005. "Obliterating Myths About Minority Institutions." *Physics Today* 58(9):49-53. https://doi.org/10.1063/1.2117823.

Sakulku, J. 2011. "The Impostor Phenomenon." *International Journal of Behavior Science* 6(1):75-97. https://doi.org/10.14456/ijbs.2011.6.

Salazar, M.R., T. K. Lant, S.M. Fiore, and E. Salas. 2012. "Facilitating Innovation in Diverse Science Teams Through Integrative Capacity." *Small Group Research* 43(5):527-558. https://doi.org/10.1177%2F1046496412453622.

Sato, S., P.M. Gygax, J. Randall, and M. Schmid Mast. 2021. "The Leaky Pipeline in Research Grant Peer Review and Funding Decisions: Challenges and Future Directions." *Higher Education* 82(1):145-162. https://doi.org/10.1007/s10734-020-00626-y.

Schutz, W. 1994. *The Human Element: Productivity, Self-Esteem and the Bottom*. Jossey-Bass, San Francisco, CA.

Seymour, E.S., and N.M. Hewitt. 1997. *Talking About Leaving: Why Undergraduates Leave the Sciences*. Westview Press, Boulder, CO.

Shapin, S. 2008. *The Scientific Life: A Moral History of a Late Modern Vocation*. University of Chicago Press, Chicago, IL.

Sheltzer, J.M., and J.C. Smith. 2014. "Elite Male Faculty in the Life Sciences Employ Fewer Women." *Proceedings of the National Academy of Sciences* 111(28):10107-10112. http://www.pnas.org/content/111/28/10107.full.pdf.

Slaton, A.E. 2010. *Race, Rigor, and Selectivity in U.S. Engineering: The History of an Occupational Color Line*. Harvard University Press, Cambridge, MA.

Smith, J.L., K.L. Lewis, L. Hawthorne, and S.D. Hodges. 2013. "When Trying Hard Isn't Natural: Women's Belonging with and Motivation for Male-Dominated STEM Fields as a Function of Effort Expenditure Concerns." *Personality and Social Psychology Bulletin* 39:3-15. https://doi.org/10.1177%2F0146167212468332.

Sommers, S.R. 2006. "On Racial Diversity and Group Decision Making: Identifying Multiple Effects of Racial Composition on Jury Deliberations." *Journal of Personality and Social Psychology* 90(4):597-612. https://psycnet.apa.org/doi/10.1037/0022-3514.90.4.597.

Stainback, K., D. Tomaskovic-Devey, and S. Skaggs. 2010. "Organizational Approaches to Inequality: Inertia, Relative Power, and Environments." *Annual Review of Sociology* 36:225-247. https://doi.org/10.1146/annurev-soc-070308-120014.

Stassun, K.G., S. Sturm, K. Holley-Bockelmann, A. Burger, D. Ernst, and D. Webb. 2011. "The Fisk-Vanderbilt Master's-to-Ph.D. Bridge Program: Recognizing, Enlisting, and Cultivating Unrealized or Unrecognized Potential in Underrepresented Minority Students." *American Journal of Physics* 79:374. https://doi.org/10.1119/1.3546069.

Stets, J., P. Brenner, P. Burke, and R. Serpe. 2016. "The Science Identity and Entering a Science Occupation." *Social Science Research* 64:1-14. https://doi.org/10.1016/j.ssresearch.2016.10.016.

Stokols, D., S. Misra, R.P. Moser, K.L. Hall, and B.K. Taylor. 2008. "The Ecology of Team Science—Understanding Contextual Influences on Transdisciplinary Collaboration." *American Journal of Preventive Medicine* 35(2):S96-S115. https://doi.org/10.1016/j.amepre.2008.05.003.

Strolger, L., and N. Natarajan. 2019. "Doling Out Hubble Time with Dual-Anonymous Evaluation." *Physics Today*. https://physicstoday.scitation.org/do/10.1063/PT.6.3.20190301a/full.

Tan, E., H. Kang, T. O'Neill, and A. Calabrese Barton. 2013. "Desiring a Career in STEM-Related Fields: How Middle School Girls Articulate and Negotiate Identities-in-Practice in Science." *Journal of Research in Science Teaching* 50(10):1143-1179. https://doi.org/10.1002/tea.21123.

Tao, K.W., and A.M. Gloria. 2019. "Should I Stay or Should I Go? The Role of Impostorism in STEM Persistence." *Psychology of Women Quarterly* 43(2):151-164. https://doi.org/10.1177%2F0361684318802333.

Thiry, H., T.J. Weston, R.P. Harper, D.G. Holland, A.K. Koch, and B.M. Drake. 2019. *Talking About Leaving Revisited*. Springer, Boston, MA.

Ting-Toomey, S., and J.G. Oetzel. 2001. *Managing Intercultural Conflict Effectively*. Sage Publishing, Los Angeles, CA.

Urban, W.J. 2010. *More Than Science and Sputnik: The National Defense Education Act of 1958*. The University of Alabama Press, Tuscaloosa, AL.

Valantine, H.A., and F.S. Collins. 2015. "National Institutes of Health Addresses the Science of Diversity." *Proceedings of the National Academy of Sciences* 112(40):12240-12242. https://doi.org/10.1073/pnas.1515612112.

Vertesi, J. 2020. *Shaping Science: Organizations, Decisions, and Culture of NASA's Teams*. University of Chicago Press, Chicago, IL.

Vertesi, J. 2021. "Shaping Science: Organizations, Decisions, and Culture on NASA's Teams." Presentation to the National Academies Committee on Increasing Diversity and Inclusion in the Leadership of Competed Space Missions on April 27.

Vincent-Ruz, P., and C.D. Schunn. 2018. "The Nature of Science Identity and Its Role as the Driver of the Student Choices." *International Journal of STEM Education* 5(48).

Waisbren, S.E., H. Bowles, T. Hasan, K.H. Zou, S.J. Emans, C. Goldberg, Sandra Gould, et al. 2008. "Gender Differences in Research Grant Applications and Funding Outcomes for Medical School Faculty." *Journal of Women's Health* 17(2):207-214. https://doi.org/10.1089/jwh.2007.0412.

Wessen, R., C.S. Borden, J.K. Ziemer, R.C. Moeller, J. Ervin, and J. Lang. 2013. "Space Mission Concept Development Using Concept Maturity Levels." Meeting paper for the AIAA SPACE 2013 Conference and Exposition. California Institute of Technology, Pasadena, CA. https://trs.jpl.nasa.gov/bitstream/handle/2014/44299/13-3547_A1b.pdf?sequence=1.

White, A.M., J.T. DeCuir-Gunby, and S. Kim. 2019. "A Mixed Methods Exploration of the Relationships Between the Racial Identity, Science Identity, Science Self-Efficacy, and Science Achievement of African American Students at HBCUs." *Contemporary Educational Psychology* 57:54-71. https://doi.org/10.1016/j.cedpsych.2018.11.006.

White, S., G. Anderson, and R. Ivie. 2011a. *Results from the 2011 Survey of Solar, Space & Upper Atmospheric Physicists.* American Institute of Physics, Statistical Research Center, College Park, MD. http://dx.doi.org/10.7302/25.

White, S., R.Y. Chu, and R. Ivie. 2011b. *2011 Survey of the Planetary Science Workforce.* American Institute of Physics, Statistical Research Center, College Park, MD. https://lasp.colorado.edu/home/mop/resources/planetary-science-workforce-survey/2011-planetary-science-survey.

Wu, Y. 2010. "Tackling Undue Concentration of Federal Research Funding: An Empirical Assessment on NSF's Experimental Program to Stimulate Competitive Research (EPSCoR)." *Research Policy* 39(6):835-841. https://ideas.repec.org/a/eee/respol/v39y2010i6p835-841.html.

Xu, C., and R.E. Lastrapes. 2021. "Impact of STEM Sense of Belonging on Career Interest: The Role of STEM Attitudes." *Journal of Career Development* (July). https://doi.org/10.1177%2F08948453211033025.

Zurbuchen, T.H. 2019. "Writing Successful Proposals: Observations from NASA." Presentation to the National Academies Committee on Increasing Diversity and Inclusion in the Leadership of Competed Space Missions on June 5. NASA Science Mission Directorate. https://smd-prod.s3.amazonaws.com/science-red/s3fs-public/atoms/files/06-05-2019_Zurbuchen_Writing%20Successful%20Proposals_FINAL.pdf.

# Appendixes

# A

# Acronyms and Abbreviations

| | |
|---|---|
| AAMC | American Association of Medical Colleges |
| AAS | American Astronomical Society |
| AAUP | American Association of University Professors |
| ADVANCE | Organizational Change for Gender Equity in STEM Academic Professions |
| AGU | American Geophysical Union |
| AHN | Native American/Alaskan/Hawaiian |
| AIP | American Institute of Physics |
| AO | announcement of opportunity |
| ARC | Ames Research Center |
| | |
| B&P | bid and proposal |
| BOSE | Board on Science Education |
| BPC | Broadening Participation in Computing |
| BUILD | Building Infrastructure Leading to Diversity |
| | |
| CDR | Critical Design Review |
| CEOSE | Committee on Equal Opportunities in Science and Engineering |
| CI | Cooperative Institute |
| CISE | Computer and Information Science and Engineering |
| CML | concept maturity level |
| Co-I | co-investigator |
| CSR | Concept Study Report |
| CV | curriculum vitae |
| CYGNSS | Cyclone Global Navigation Satellite System |
| | |
| DAPR | Dual Anonymous Peer Review |
| DEI | diversity, equity, and inclusion |
| DEIA | diversity, equity, inclusion, and accessibility |
| DMR | Division of Materials Research |

DoD Department of Defense
DOE Department of Energy
DPS Division of Planetary Science

EPP Educational Partnership Program
EV Earth Venture
EV-C Earth Venture Continuity
EV-I Earth Venture Instrument
EV-M Earth Venture Mission
EV-S Earth Venture Suborbital

FAR Federal Acquisitions Regulation
FFRDC federally funded research and development center
FINESST Future Investigators in NASA Earth and Space Science and Technology
FUSE Far Ultraviolet Spectroscopic Explorer

GOES Geostationary Operational Environmental Satellites
GOLD Geosciences Opportunity for Leadership in Diversity
GRAIL Gravity Recovery and Interior Laboratory
GSFC Goddard Space Flight Center

HBCU Historically Black College or University
HBCU-EiR Historically Black Colleges and Universities Excellence in Research
HEOMD Human Exploration and Operations Mission Directorate
HQ Headquarters
HSI Hispanic-Serving Institution

IDeA Institutional Development Award
IDS Interdisciplinary Research Science
INBRE IdeA Networks of Biomedical Research Excellence
INCLUDES Inclusion across the Nation of Communities of Learners of Underrepresented Discoverers in Engineering and Sciences
IRACDA Institutional Research and Academic Career Development Awards
IRAD Internal Research and Development
IRG Interdisciplinary Research Group
ITAR International Traffic in Arms Regulations

JHU/APL Johns Hopkins University Applied Physics Laboratory
JPL Jet Propulsion Laboratory

LaRC Langley Research Center
LASP Laboratory for Atmospheric and Space Physics (University of Colorado Boulder)
LGBTQ+ lesbian, gay, bisexual, transgender, queer
LPSC Lunar and Planetary Science Conference

MARC Maximizing Access to Research Careers
MDAA Mission Directorate Associate Administrator
MDR Mission Definition Review
MIDEX Medium-Class Explorer
MIRO MUREP Institutional Research Opportunity

| | |
|---|---|
| MoO | mission of opportunity (also MO) |
| MOSAIC | Maximizing Opportunities for Scientific and Independent Careers |
| MRSEC | Materials Research Science and Engineering Center |
| MSI | Minority Serving Institution |
| MUCERPI | Minority University and College Education and Research Partnership Initiative |
| MUREP | Minority University Research and Education Programs |
| | |
| NAC | NASA Advisory Council |
| NARCH | Native American Research Centers for Health |
| NASA | National Aeronautics and Space Administration |
| NCAS | NASA Community College Aerospace Scholars |
| NCES | National Center for Education Statistics |
| NEAR | Near Earth Asteroid Rendezvous |
| NEC | NIH Equity Committee |
| NHLBI | National Heart, Lung, and Blood Institute |
| NIGMS | National Institute of General Medical Sciences |
| NIH | National Institutes of Health |
| NOAA | National Oceanic and Atmospheric Administration |
| NRC | National Research Council |
| NRMN | NIH National Research Mentoring Network |
| NSF | National Science Foundation |
| NSPIRES | NASA Solicitation and Proposal Integrated Review and Evaluation System |
| | |
| ODEO | Office of Diversity and Equal Opportunity |
| OECR | Office of Equity and Civil Rights |
| OPM | Office of Personnel Management |
| OSTEM | Office of STEM Engagement |
| OSTP | Office of Science and Technology Policy |
| | |
| PDR | Preliminary Design Review |
| PEA | Program Element Appendix |
| PhD | doctor of philosophy |
| PI | principal investigator |
| PM | project manager |
| PREM | Partnerships for Research and Education and Materials |
| PSE | project systems engineer |
| | |
| R&A | research and analysis |
| R&D | research and development |
| RaMP | Research and Mentoring for Postbaccalaureates in Biological Sciences |
| REM | Research Experience and Mentoring |
| RENEW | Reaching a New Energy Sciences Workforce |
| RFI | request for information |
| RFP | request for proposal |
| ROSES | Research Opportunities in Space and Earth Sciences |
| | |
| S&E | science and engineering |
| SAC | STEM Advisory Council |
| SALMON | Stand-Alone Missions of Opportunity |
| SAMPEX | Solar, Anomalous, and Magnetospheric Particle Explorer |

| | |
|---|---|
| SIMPLEX | Small Innovative Missions for Planetary Exploration |
| SMART | Space Mission Architecture and Risk Analysis Tool |
| SMD | Science Mission Directorate (NASA) |
| SMD AA | Associate Administrator for the Science Mission Directorate |
| SMEX | Small Explorer |
| SOMA | Science Office for Mission Assessments |
| SSB | Space Studies Board |
| STEM | science, technology, engineering, and mathematics |
| STMD | Space Technology Mission Directorate |
| SWAS | Submillimeter Wave Astronomy Satellite |
| SWRI | Southwest Research Institute |
| | |
| T&C | term and condition |
| TCU | Tribal College and University |
| TMC | technical, management, and cost |
| TRL | technology readiness level |
| | |
| URM | underrepresented minority |
| | |
| WBS | work breakdown structure |
| W/M | woman/man |

# B

# Glossary

Accessibility—The design of products, devices, services, or environments so that people with disabilities can fully participate, engage, and benefit.

Authentic research—"Real" research that involves direct engagement with the actual work of missions or that provides training that is directly relevant to real mission work.

Best practice—"A procedure that has been shown by research and experience to produce optimal results and that is established or proposed as a standard suitable for widespread adoption" (reference in Chapter 6, footnote 1)

Bias—Attitudes, behaviors, and actions that are prejudiced in factor of or against one person or group compared to another.

Disciplinary climate—How individuals experiences their field, institution or department; a manifestation of culture.

Disciplinary culture—The implicit or explicit customs, behaviors, norms, and values that are deeply embedded within a discipline or field.

Diversity—The practice or state of involvement of persons across a variety of social and demographic characteristics within organizational, institutional, and interactional settings.

Emerging practices—Best practices that are new, innovative, or exploratory in nature. They may be based on some level of evidence, but it is not sufficient for the practices to be considered "promising" or "evidence-based."

Equity—The absence of barriers, biases, and obstacles that impede access, fair treatment, and opportunity for contribution by all members of a community, recognizing that different resources or approaches are needed to remedy the uneven playing field that exists across different groups.

Ethnicity—Socially constructed and historically contingent categorizations of humans based on perceptions of shared culture, such as language, ancestry, practices, and beliefs.

Evidence-based practices—Best practices that are grounded in research and evaluation and have met some established test of validity in improving a specific outcome.

Explicit bias—Conscious, intentional bias.

Gender identity—The way one identifies oneself within or outside of traditional, socially constructed categories of women and men. Gender identity is separate from biological sex (which is a socially constructed dichotomy of bodies based on perceptions of shared physical and/or chromosomal traits), and may include gender non-binary and gender queer identities that lie outside of the traditional gender binary.

Historically minoritized communities in STEM—Socio-demographic groups (e.g., certain genders, racial/ethnic groups) that, as a result of historical and contemporary processes of oppression and bias, have been excluded from full and representative participation in science, technology, engineering, and medicine (STEM).

Homophily—Referring to people seeking out persons similar to themselves.

Implicit bias—A type of "unconscious" bias that occurs automatically and unintentionally, but affects an individual or group's judgments, behaviors, and decisions.

Imposter phenomenon—The psychological experience of professional or intellectual fraudulence, disproportionately affecting historically underrepresented groups.

Inclusion—The practice of facilitating the equal distribution of opportunities, resources, and recognition to persons regardless of social and demographic characteristics within organizational, institutional, and/or interactional environments.

Institutional/organizational—Describes the conditions of a professional or academic environment, including availability of resources and opportunities, implicit and explicit disciplinary norms, values and institutional culture, and institutional and agency practices and policies.

Interpersonal—Describes relational factors, including interactions between individuals and groups shaped by social and personal identity, which can have a differential impact on historically underrepresented groups.

Intersectionality—A critical, theoretical, and analytical framework that highlights how multiple social identities such as race, gender, sexual orientation, socioeconomic status, and disability intersect at the micro level of individual experience to reveal interlocking systems of privilege and oppression (i.e., racism, sexism, heterosexism, classism) at the macro social-structural level.

Intrapersonal—Describes beliefs, biases, and thought processes that shape decision-making, the judgments we make about others, and how we view ourselves.

Promising practices—Best practices that have been successfully implemented, but sufficient evidence that substantiates all the parameters associated with success of the practice has not been collected or generated.

Race—Socially constructed and historically contingent categorizations of humans based on perceptions of shared physical and/or social traits.

Sexual identity—The way one identifies oneself in terms of to whom one is romantically or sexually attracted. May be within or outside of traditional, socially constructed categories of heterosexuality and non-heterosexuality, and may include bisexual, queer, fluid, and asexual identities.

Social ecological model—Developed by Urie Bronfenbrenner to understand human development. Consists of concentric circles of influence and networks, applied to describe a broad range of social processes.

Structural—Describes the systemic factors that originate from sociohistorical and sociocultural conditions and act to constrain access to power, resources, and opportunity on the basis of identity, resulting in advantages for some groups and disadvantages for others.

Tokenism—The act of making only symbolic efforts to increase representation of persons from minoritized groups, particularly by recruiting one or two people from those groups to give the appearance of equality within an organization or work group.

# C

# Qualitative Study of the Factors That Impede or Facilitate Diversity, Equity, Inclusion, and Accessibility in the Proposed Leadership for Competed Space Missions


Prepared for

National Academies of Sciences, Engineering, and Medicine
Committee on Increasing Diversity and Inclusion in the Leadership of Competed Space Missions
500 Fifth Street. NW
Washington, DC 20001

Prepared by

Mithuna Srinivasan, PhD
Heather Sawyer, PhD
Ann Kearns Davoren, PhD
Julie Kubelka, MSEd
Elena Navarro, BA
NORC at the University of Chicago
1155 East 60th Street, 2nd Floor
Chicago, IL 60637


## INTRODUCTION

### Background to the Study

Driven by concern over a lack of diversity in the pool of applicants for competed space missions, the National Aeronautics and Space Administration (hereafter, NASA) commissioned the National Academies of Sciences, Engineering, and Medicine (hereafter, the National Academies) to conduct a consensus study aimed at understanding the factors that impede or facilitate diversity, equity, inclusion, and accessibility (DEIA) in the proposed leadership for competed space missions supported by the Science Mission Directorate (SMD) at NASA.

The consensus study is being conducted by the *Committee on Increasing Diversity and Inclusion in the Leadership of Competed Space Missions* (hereafter, the Committee) at the National Academies. The Committee's key charge is to recommend actions to increase DEIA in the leadership of space mission proposals submitted to the NASA SMD competed space mission programs. For this, the Committee is engaging with a wide range of experts in the relevant social science and space science communities to achieve the following objectives:

- Map the current space mission proposal system
- Identify elements of the system that may present barriers or bottlenecks to potential proposers, yielding a less diverse pool of proposal leadership teams
- In the step above, identify humanistic elements of the system that may present impediments to applicants, limiting the diversity of the competitive pool
- Use existing NASA proposer data and other sources to analyze differences between the pool of current and past SMD proposers and the community of researchers in NASA-relevant fields, the larger U.S. science community, and the national population
- In developing recommendations, consider best practices from other agencies funding large, principal investigator (PI)-led, multi-institutional projects to find common bottlenecks and to identify innovative methods for overcoming such barriers
- Recommend practical and effective actions that NASA SMD and its partners can execute to help proposers overcome barriers to diversity and inclusion to ensure a diverse pool of proposal leadership teams
- Recommend ranges of activities and choices that individuals may make during their career to enhance the likelihood of individual success as a future leader of a competed space mission

## Research Questions

To support the work of the Committee in addressing the above objectives, in June 2021 the National Academies contracted with NORC at the University of Chicago (hereafter, NORC) to organize and conduct **virtual, in-depth interviews with individuals who have prepared and submitted at least one proposal to the NASA SMD's competed space missions program as the PI from 2010 to the present.** The purpose of the interviews was to gather information about applicants' experiences with the NASA SMD mission proposal development and submission processes, as well as perceived barriers and opportunities to ensuring a diverse, competitive pool of proposed leaders. As such, insights gleaned from the interviews will help the Committee address the first three objectives listed above. Additionally, the interviews will inform practical recommendations that the Committee can make for NASA SMD and its partners to reduce barriers and bottlenecks in the current mission proposal process in order to yield a more diverse pool of proposal leadership teams.

Against this backdrop, the current study addresses the following **research questions:**

- What educational and earlier career experiences led competed space mission PIs to consider submitting to NASA?
- How did competed space mission PIs learn about funding opportunities, how were they supported/mentored (or not) through the lead-up to and actual proposal submission process, and what factors helped or hindered their decision to apply?
- What were the experiences of competed space mission PIs from the proposal submission stage forward (e.g., during site visits, evaluations, or moving forward after receiving denials)?

## Report Purpose and Structure

This report presents findings from the qualitative data collection with competed space mission PIs. The findings are organized by the three research questions listed above and aim to provide an understanding of individuals' educational and career experiences that shaped their decision to pursue competitive funding from NASA, the availability of key facilitators that helped individuals realize their full research potential and successfully make grant submissions to NASA, challenges that they faced along the way to proposal submission and after, and their overall assessment of the proposal submission process and outcomes at NASA.

The rest of the report is organized as follows. Chapter 2 discusses our methodology, including details on respondent outreach, the sampling and recruitment of interview participants, data collection and analysis, and study limitations. Chapter 3 describes the key findings from the data collection, organized by the three research questions. Chapter 4 concludes with a high-level recapitulation of the study's methodology and key findings.

## METHODOLOGICAL APPROACH AND DATA COLLECTION

As previously noted, the key aim of the study was to conduct individual, semi-structured interviews with scientists who, between 2010 and present, had prepared and submitted at least one space mission proposal to the NASA SMD as a PI. The study received approval from the National Academies' Institutional Review Board (IRB).

### Eligible Respondent Identification

Scientists with experience preparing and submitting at least one space mission proposal as the PI in response to an announcement of opportunity (AO) from the NASA SMD from 2010 to the present were considered eligible for study participation.

In order to identify individuals eligible to participate in the study, the NORC team, National Academies project staff, and members of the Committee jointly conducted email outreach to various space science community groups and organizations with a request for them to disseminate within their networks information about the study and a web-based Qualtrics Study Eligibility Survey. In order to more directly target the NASA PI proposer pool for competed missions, the same community recruitment email and link to the Study Eligibility Survey also was sent to the NASA SMD with a request for them to circulate it to their proposer listserv/database. Appendix A presents the full list of organizations and groups that were contacted by the NORC team, National Academies project staff, and members of the Committee; Appendix B contains the community outreach emails; and Appendix C presents the Study Eligibility Survey.

Interested individuals self-selected into the study by voluntarily completing the Study Eligibility Survey, indicating on that survey if they would consent to be contacted for an interview and accordingly providing their contact information as part of the survey (see Exhibit 1). Respondents also received a notification before exiting the survey that they will be contacted within a few weeks for an interview by the NORC team if deemed eligible. The survey also noted that, if the respondent did not receive this follow-up from NORC by a specified deadline, they could assume that they were not selected. Survey respondents also were asked to share the survey link with others who they thought might be interested in participating in the study.

### Selection and Recruitment of Interview Participants

The Study Eligibility Survey was administered from August 3-25, 2021. During this period, 320 participants completed the survey. Of them, 64 were considered eligible for participation in the study. The NORC team selected a subset of 40 eligible individuals to invite for interviews, with the aim of maximizing diversity across background characteristics and proposal submission history. All 40 individuals received a personal email outreach from the NORC team inviting them for a one-hour virtual interview at a date and time convenient for the respondent. Appendix D contains a copy of the individual outreach email. NORC sent one follow-up reminder by email if the individual did not respond within three days. If still no response was received, the individual was dropped from consideration.

The selection of the 40 individuals for interview outreach was done in a staggered fashion across two rounds:[1]

[1] The wording of survey question 1 changed over the period of the survey administration, which is why interviewee selection was done in two rounds. In the version of the survey that was used during Round 1, question 1 appeared as: *Have you ever served as principal investigator (PI) on a proposal for a competed space mission submitted to NASA SMD from 2010 to the present?* This question did not distinguish between those who had been PIs on instrument proposals versus full space mission proposals. Per the National Academies' guidance, the NORC team subsequently modified this question for Round 2 to appear as: *Have you ever served as principal investigator (PI) on a proposal for a competed space mission submitted to NASA SMD from 2010 to the present (excluding instrument PIs)?*

**Exhibit 1:** Study Eligibility Survey

The first question of the survey was a **screening question that asked about PI status**; individuals who had never been a PI on a NASA SMD proposal for a competed space mission received a real-time notification of study ineligibility and exited the survey.

▪Eligible participants were then asked to **indicate their interest in being contacted for a confidential interview**; individuals not interested received a real-time notification to exit the survey.

Interested individuals were asked to **respond to the remaining questions in the survey** that asked about background characteristics (gender identity, race/ethnicity, career stage, professional affiliation, and geographic location of work) and proposal submission history (mission proposal funding status, NASA space mission divisions from which funding was sought, and the size of mission proposal budgets).

Finally, they answered **questions seeking contact information** for a potential follow-up by the NORC team for an interview.

- Round 1 outreach was based on responses to the survey that were received August 3-17, 2021, which resulted in 12 eligible respondents. No sampling was done in this round; instead, the NORC team selected all 12 for interview outreach. **Interviews were scheduled with 11 of the 12 invitees during the Round 1 outreach**.
- Round 2 outreach was based on responses to the survey that were received August 18-25, 2021, which resulted in 52 eligible respondents. To reach a targeted pool of 40 individuals for interview outreach, NORC purposively sampled 29 eligible individuals from Round 2 due to only 11 of the 12 Round 1 respondents participating. The following section presents details of the purposive sampling that was done for Round 2. **Interviews were scheduled with 18 of the 29 invitees during the Round 2 outreach**.

### Characteristics of Interviewee Pool

As mentioned above, in total there were 64 **eligible survey respondents.** The majority were White (67%), while 17% identified as Other/Multiracial, 11% as Asian/Asian-American, and 5% as African-American/Black. With respect to ethnicity, 6% identified as being Hispanic/Latinx. The majority of eligible survey respondents also identified as men (73%), were professionally affiliated with an academic institution (56%) and were from the Western United States (65%). With regard to their proposal submission history, for the majority (80%) their most recent bid to NASA was unsuccessful, and the mission size of their most recent bid was very small, at under $125 million (58%). Mid-career professionals (with 11-25 years of experience) made up 44% of the eligible respondents, early career professionals (0-10 years of experience) comprised 20% of eligible survey respondents, and late career (more than 25 years of experience) made up 25% of eligible survey respondents. Finally, respondents were asked to what space mission division they have applied for funding throughout their careers. Planetary Science comprised the largest proportion at 25%, and Earth Science was selected by 23% of eligible survey respondents.

Nearly 20% of eligible survey respondents had applied to Astrophysics, 9% to Heliophysics, and an additional 23% had applied to multiple divisions.

As previously mentioned, all 12 eligible individuals from Round 1 were invited for an interview, and 11 of those 12 participated. With 52 eligible participants in Round 2, the NORC team implemented a three-step purposive sampling strategy to select 29 individuals to invite for interviews for this round. The purposive sampling was done in such a way as to include an adequate number of respondents from historically marginalized groups in STEM fields. Therefore, the first step sampled by race and included 17 non-White eligible respondents; the second step sampled by gender identity and included eight additional respondents who identified as women; and the third step sampled by space mission division and included four individuals who at some point in their career submitted a proposal to the Heliophysics division.

The result was an invited interview pool of 40 individuals that maximized diversity among key characteristics. **Among those invited for an interview:**

- 40% reported they were non-White or multiracial
- 40% self-identified as women
- An almost even split was achieved between successful and unsuccessful most recent bids
- 28% had applied to the Planetary Science division, and 25% to the Earth Science division
- 70% reported an academic affiliation[2]

Of the 40 invited for an interview, discussions were scheduled with 29 respondents who accepted the invitation. In the **final interviewee pool,** the majority ended up being White (69%), while 52% of participants identified as women. Most (72%) had an academic institution professional affiliation, and most were from the Western United States (59%). Most participants were either early- (24%) or mid-career (48%) and had been unsuccessful in getting their most recent bid as PI to NASA funded (66%). Nearly 60% of interview participants submitted very small bids as part of their most recent submissions to NASA. Planetary Science was the most frequently occurring space mission division (38% selected this), and an additional 21% had applied to Earth Science.

## Data Collection

**Instrument development.** The NORC team engaged in formative research to gain insights needed for the development of an interview guide that would meet the objectives of this study. We reviewed the following background material provided by the National Academies:

- The *Explore Solar System and Beyond: NASA Announcement of Opportunity Science Team Demographics presentation* given by Heidi Jensen and Lorenzo Pappas, June 2021
- The consensus study report *Sexual Harassment of Women: Climate, Culture, and Consequences in Academic Sciences, Engineering, and Medicine* from the National Academies, 2018
- The PI-Led Mission Proposal Process presentation given by Dr. Thomas Wagner from the Planetary Science Division at NASA, February 2021
- The *Goddard Space Flight Center Proposal Development* presentation given by Deborah Amato from the NASA Goddard Space Flight Center, February 2021
- The *Proposal Process: My Perspective from Planetary Discovery and New Frontiers Mission Proposals* presentation given by Cathy Olkin from the Southwest Research Institute (SwRI), February 2021
- The *Writing Successful Proposals: Observations From NASA* presentation given by Dr. Thomas Zurbuchen from the NASA SMD
- The *Being a First Time PI* presentation given by Dr. Erika Hamden, February 2021.

[2] Of eligible respondents, 56% reported an academic institution affiliation. Our invited interview pool was more heavily skewed toward academics because of prioritizing other areas for sample, including race/ethnicity, gender identity, and space mission division.

- The *SMD Summary Demographic Data* presentation given by Louis Barbier, Caroline Wilson, and Ankita Kc from the Office of the Chief Scientist at NASA, June 2021.

After reviewing the above materials, the NORC team prepared a semi-structured interview guide to conduct interviews with selected individuals. Exhibit 2 presents the sections and main topics covered in the interview guide. A full copy of the interview guide is available in Appendix E.

**Interview process.** Semi-structured interviews were conducted by the NORC team virtually using the Zoom platform from August 25-September 22, 2021. Each interview lasted up to an hour and was audio-recorded with the respondent's consent. Each discussion was facilitated by two NORC project staff, with one serving as the interviewer and the other as the notetaker. Interviewers and notetakers used a written protocol that included clear and concise instructions as well as suggested probes to help guide conversations with participants.

Prior to starting the interview, the interviewer verbally reviewed consent protocols with each participant to ensure they understood the purpose, risks, and benefits of their participation. The notetaker documented in the interview notes whether verbal consent was granted. The participant was also told that, following the interview, they could receive an electronic copy of the consent form if they would like one, and the notetaker documented in the interview notes whether the participant opted to receive a follow-up email with the consent form. The consent to record was also documented in the interview notes.

## Data Analysis

Each interview was transcribed manually by a member of the NORC team. Analysis of early interview transcripts was done using a modified grounded theory approach, whereby emerging themes were used to build explanatory models of relationships between items. When possible, the team implemented "in vivo" coding, which used respondent's own words and phrases as themes, such as PIs needing to be able to "speak multiple languages" across proposal teams. This enabled the team to generate probes iteratively from interview to interview, to incorporate respondent feedback on topics.

Throughout the data collection period, the NORC team met often to discuss emerging findings and kept a running list of themes, which was expanded upon as new topics emerged until saturation occurred. Preliminary themes and findings were shared with National Academies project staff and the Committee to determine their relevance and potential inclusion in the analysis. After all the interviews concluded, the team performed deductive coding and analysis on the entirety of interview findings, using the preliminary themes that had been identified previously as well as any new themes that emerged. A flat coding structure was used to assign equal relevance to all themes. Each team member in charge of the analysis was assigned a subset of interviews to review to ensure the comprehensiveness of themes and perform coding verification. After coding and analysis were complete, overlapping thematic topics were consolidated and organized by the three research questions.

**Exhibit 2:** Interview Guide Layout

| Section | Topics Covered |
|---|---|
| Participant Career Background and Experiences | • Early exposure to NASA competed missions<br>• Types of incentives or supports received from employers<br>• Mentoring-related experiences |
| Lead-Up Process to Proposal Submission | • Process of forming mission teams<br>• Barriers and facilitators to proposal development and submission |
| Post-Submission Process and Outcomes | • Post-submission processes<br>• Challenges faced post submission<br>• Experiences of being a PI |

### Limitations of the Research

This study provides a snapshot of the experiences of proposed PIs for competed space missions in their career background, proposal submission process, and post-submission process. Some limitations of this study include the following:

- Although our sample of 29 interviewees allowed us to capture rich, qualitative data with some diversity along race/ethnicity, gender identity, and space mission division, as is a constraint with all qualitative research, the generalizability of findings is limited.
- Self-reported data from interviews can suffer from issues such as recall error.
- Subgroup analyses that stratified findings by background characteristics such as race/ethnicity, career stage, subfield, and proposal submission history were not always feasible, given the risk posed to the confidentiality of participants. Therefore, experiences of PIs in subgroups may vary.
- A total of 12 respondents across Round 1 and Round 2 did not respond to our request for an interview. Therefore, findings may suffer from some degree of non-response bias. That is, the opinions of those who did not respond may be different than those who were interviewed and whose responses were included in this study.
- This study focused exclusively on individuals who have submitted at least one full mission proposal bid as PI to NASA in the last 10 years. Therefore, the study team did not systematically investigate the experiences of co-Investigators (Co-Is), Deputy PIs, post-docs, graduate students, and others who may face significant barriers while on, or trying to join, proposal teams. We were also unable to systematically examine "barriers to entry" to proposal submission. Thus, study findings should be interpreted keeping these exclusion criteria in mind, and only reflect the experiences and perceptions expressed by *select individuals who submitted at least one full mission proposal to NASA as a PI* over the course of their career.

## FINDINGS

While all respondents had earned PhDs in their respective fields, many commented that they did not learn all the skills needed to be a successful PI through their undergraduate and graduate school training. Similarly, respondents working at academic institutions noted that higher education is "woefully inadequate" in preparing young scientists to be PIs. Furthermore, the types of skills and experiences to hone that would be beneficial for career growth are not known by many students.

***Higher education opportunities set the stage for later success.*** *However, respondents explained that, while academic training focuses on providing students with a sound scientific background, they felt there is a need to broaden the skills taught in higher education. For example, most programs do not include classes on management, budgeting, public speaking, conflict resolution, or proposal writing—all skills noted to be of importance to PIs.*

Despite these limitations, conversations with most respondents about their educational experiences revealed that they felt higher education was a critical time when they gained some important skills that gave them a competitive edge over many of their peers. For example, they found mentors, started building their networks, and gained access to opportunities that provided them with the mission experience that helped them later become PIs.

Respondents gave several suggestions as to how time spent in higher education could better prepare students for PIship. One suggestion was to better align the length of time spent in school with mission frequencies. It was perceived that, due to the infrequency of large mission opportunities, many students do not have the chance to join a mission and gain mission experience. Respondents suggested that increasing the number and frequency of smaller missions would be one way to ensure that more students gain this vital experience. It was also noted that increasing the number of smaller missions would allow young postgraduates to get the necessary experience as a PI or Co-I on a smaller project, which increases their qualifications to lead a larger project. Other suggestions included requiring courses on proposal writing, budgeting, public speaking, conflict resolution, and project management to bolster the skills needed for PIship.

**Becoming a PI is in part predicated on serving in positions of increasing responsibility, so early career mission experience was perceived as being important.** Many respondents worked under advisors in undergraduate and graduate school who were working on, or had connections to those working on, NASA space missions. As students, these respondents engaged in activities such as mission data collection, participating in meetings, and serving on review panels. These experiences enabled them to learn firsthand about the mission process and potential areas of scientific exploration and helped build their careers.

Respondents who entered their scientific fields at later ages or who experienced delays in early career development reported feeling perpetually "behind" their peers in terms of publications, grants, and other professional accomplishments. A lack of formalized career steps or career development ladders (as early as the undergraduate level) was perceived to increase barriers, as the steps needed to advance a career were not always well understood. Respondents also felt there is a need for a greater number of small mission opportunities to enable early career professionals to gain experience writing proposals, managing teams, and developing innovative technology.

**Many respondents reported having strong support from a mentor, and some were even told explicitly they would make a good PI early in their careers. They felt that these mentorship experiences were foundational to their sense of belonging in the field.** Mentors introduced respondents to others in their network, connected them with mission opportunities, and helped them understand the mission proposal process. Some respondents perceived that the respect and reputation of their mentor was transferred to them as their career progressed, which helped them gain credibility and name recognition. Respondents noted that the culture of the scientific community needs to shift in such a way so that experienced PIs are encouraged to mentor a wider swath of students and early career scientists. One suggestion to address this issue was to create a NASA database to help pair potential mentors and mentees across organizations and networks.

**Insofar as networks are important sources of information sharing, some respondents perceived that NASA training opportunities are not always widely publicized and are more accessible to those who are well connected.** The scientific community working on space missions was characterized as small and seemingly impenetrable for an early career scientist outside in-crowd networks. Respondents expressed that mission teams are often comprised of people who already know or have worked with each other. It was therefore perceived that students and early career scientists are often brought onto mission teams because they are (or were) the graduate students of a PI or their PI's close friend or colleague. Respondents felt this creates networks of "in crowds" that are difficult to gain access to without facilitation by someone in the network.

***Respondents from ethnic/racial minority groups and women were more likely to report receiving very little support (or even hostility) from mentors.*** *These respondents reported a range of adverse mentorship experiences, including receiving no support at all, feeling passed over for opportunities they were qualified for, and staying in toxic advisor/advisee relationships for lack of other options.*

Women and minorities were more likely to report difficulty breaking into networks as a result of negative mentorship experiences. For example, one woman had an advisor who was known to have intimate relationships with his students, and the respondent noted that she was often teased or ridiculed by her peers under the presumption she may be involved with him as well. A person identifying as a racial/ethnic minority had an advisor who not only provided no support through their training but created roadblocks to their progress. Those who did not have supportive advisors or were not in programs that had involvement with space missions were often left on their own to figure out proposal development by trial and error. For respondents who experienced negative relationships, some found support outside formal advisor relationships from others in their field, while others found mentorship from experienced scientists at their place of work.

## PROPOSAL PREPARATION AND SUBMISSION

**PIs described the time spent in preparing proposals as extremely taxing on their personal and professional lives.** Some noted they were "not great parents" during these periods. Others said that they could only devote the time required to prepare a proposal because they did not have children, had very supportive partners who did a disproportionate amount of work around the home, or could afford to hire help (such as a personal shopper). Respondents who were late in their careers had a general sense that proposals have become longer and more involved, teams have grown larger, and that the overall demands involved with proposal submissions have increased over the past few decades.

> ***Perceptions of a "qualified PI" limit the pool of proposed leaders to a small group.*** *Some respondents felt that the perpetuation of the idea that few are qualified to serve as PI was a gatekeeping mechanism used to keep the same small group of PIs in power.*

It was felt that, for early career PIs, the time and effort needed to develop proposals can prevent them from advancing their scientific findings, presenting at conferences, and publishing manuscripts, which can ultimately hinder their professional development (such as securing tenure). Some respondents felt that they had to risk accepting roles on teams or at institutions with known cultural or interpersonal challenges just so they could get their foot in the door.

**Many respondents felt that the dedication and qualifications of a proposed PI played an important role in proposal outcomes.** Some felt a PI needs to have a combination of experience that is "virtually impossible to find" in a single person, which greatly limits the pool of proposed leadership to a small group. Prior mission experience, an understanding of the administrative and management requirements, and relevant scientific background were all perceived as skills needed to serve as a PI. Due to the complex role of a PI and significant size of funding awards, respondents felt that PIs need to be "taken seriously" and seen as "trustworthy" not only by NASA, but also by industry partners and the general public. Many respondents perceived that, in selecting proposals, NASA can be swayed by other factors such as a PI's reputation, proof of past success as a PI or Co-I, publication records, accolades and appointments, and the prestige of PhD-granting institutions. Because the proposal review process is not double-blinded, respondents felt that having resumes with respected early career experiences and working with influential, established scientists in the field meant that NASA would be more likely to view a PI's submitted proposal favorably. These perceptions are important to acknowledge because they influence how proposal teams form.

> ***Mission team formation was perceived to be shaped by institutional affiliation and hidden timelines.*** *Respondents felt that those who work at NASA centers have a better understanding of NASA reporting requirements, budget monitoring, and other administrative processes. For those working outside NASA centers, few reported having administrative support from their institutions, and a perceived lack of clarity surrounding these requirements led to difficulties navigating "double decision gates" due to differences between their institution's processes and procedures and those set by NASA.*

**Respondents cited networking as a primary means of proposal team formation, but the processes of partnership and pathways to PIship differed among respondents. Factors such as institutional affiliation and personal reputation were perceived to impact how teams were assembled as they shaped differential access to information, timelines, and processes.** Some acknowledged having PIship bestowed upon them by more senior colleagues, who acted as gatekeepers within networks. Others spent years on mission teams, committees, review panels, and other working groups before obtaining the skills or reputation needed to be viewed as a qualified PI.

> ***PIs face financial, career, and personal risks.*** *Respondents mentioned the various risks that PIs, industry partners, and NASA take during the mission proposal development and submission. On a personal level, PIs felt they risk their careers by spending years developing proposals that can take up to decades to execute without the guarantee of funding. This often requires PIs to have stable streams of income to accommodate the time spent on proposals, by continuously writing proposals, working on other grants, teaching at universities, or relying on institutional funding.*

For example, respondents working at academic institutions or centers felt that they often had to "sell" their position as PI to their own institution, along with industry and management partners. Some felt they had to work very hard to be taken seriously and grow their networks, implementing strategies such as cold-calling potential partners. Several people reported using conferences as places to learn who the key power brokers were in industry and approach potential partners. Others reported difficulties partnering with mission management organizations without specific institutional affiliation or a known reputation. Some felt they had to compete against other mission teams to secure partnerships with management and industry partners.

Respondents working at NASA centers were more likely to feel that the selection of a PI was an internally competitive process because mission concepts had institutional support and the processes of partnership were formalized. Some respondents perceived a lack of clarity in training and proposal requirements to partner with industry, creating a climate of "hidden timelines" and frustration in building mission teams. Because teams form anywhere from six months to several years before an AO is released, informal networks are heavily relied on for information sharing.

**Respondents felt that, from mission conception to proposal submission, the burden of responsibility ultimately falls on the PI. Even if a proposal makes it through the first phase of review, some respondents mentioned that NASA often asks for budget cuts that force PIs to make difficult decisions about how to proceed. PIs perceived that if they deny budget cuts, then they risk losing funding in the next round.** If the PI agrees but the work cannot be completed within that new budget, it creates a range of challenges for the PI to navigate with project managers, scientists, and engineers. This can also lead to situations where PIs have to renegotiate with industry partners to prevent them from backing out.

Given the vast financial scale of space missions, many perceived NASA as being more "conservative" with risk taking regarding PI selection, proposal budgets, and industry partners. Some felt that newer, "unproven" PIs and more "innovative" science proposals may be less likely to be selected due to NASA perhaps not wanting to risk wasting taxpayer money on potentially failed projects. Respondents also felt that PIs who have prior working relationships with NASA program managers and certain industry partners seem to be more likely to win proposals. The perceived bias toward experienced PIs creates challenges for newer PIs or PIs who want to further explore emerging science areas.

## POST-SUBMISSION OUTCOMES AND MOVING FORWARD

**Site visits were perceived to have grown increasingly demanding in resources, time, and money.** Site visits were described in various ways, such as the "talk of your life," a "dog and pony show," and "like a thesis defense times 10 crossed with a Hollywood production." Participants felt that the requirements to execute a successful site visit have grown increasingly elaborate over the years. Respondents reported their teams spent weeks or months preparing for site visits, including things such as media training, dress rehearsals, cross-country travel, building exhibits, and creating project websites. These preparations put strains on the proposed PI's career responsibilities and other personal commitments (such as raising children). Overall, these events were described as resource-intensive, pressure-filled, and requiring great planning and theatrics. Participants noted that, while the costs of missions are well documented, an immense amount of capital and labor goes into simply competing for a bid.

> ***Proposal outcomes were perceived to lack consistency and clarity.*** *While a few respondents said debriefings following a proposal rejection helped them understand the reasons for not being selected, many said they felt the official reasons given by NASA did not fully reflect behind-the-scenes decision-making. Several perceived a lack of clarity in proposal selection processes and variation in the quality of debriefing feedback as contributing to these feelings.*

**Although some respondents said receiving a rejection was the most important thing that happened to their careers because they gained valuable experience and exposure, others felt that losing a bid was akin to a "death in the family" and emotionally and professionally devastating.** One respondent reported feeling used by a mission team, who they later came to feel put them in the PI role as a "token" to appease NASA's push for diversity. They later learned that key team members felt they possessed "inside information" that winning the bid was a long-shot due to another proposal being favored by NASA even before it was submitted. Another respondent who felt used as a diversity token questioned if they would ever submit another proposal after the immense demands of the preparation process and perceived unfairness of the proposal outcome. Several people reported the perception that there is a lack of clarity in which proposals get funded and which get rejected, and several felt that favoritism came into play.

> ***Participants perceived cultural and climatic challenges of being a PI.*** *Respondents of varying demographics felt that the scientific community's views of a PI's leadership capabilities often have a gender bias, as male PIs who seek opportunities are more likely to be viewed as being "confident," whereas women PIs or those identifying as racial/ethnic minorities may get the reputation of being "difficult" or "arrogant."*

**A perception that was repeated by multiple respondents was that a PI needs to be able to "speak multiple languages" to convey information effectively across diverse actors, including engineers, scientists, management, leadership at NASA, industry partners, and the general public.** Respondents felt that each of these groups has their own culture, lingo, and style of communication. Effective leadership was perceived as requiring a depth and breadth of educational, career, and life experience. Many agreed that a PI needs to foster a healthy team dynamic, although respondents reported having witnessed PIs with a range of working styles—some great and others terrible. Some of these negative traits mentioned included being domineering, not consulting the

> ***Respondents who identified as women and/or belonging to racial/ethnic minorities shared a range of troubling experiences as PIs.*** *For example, being the target of "sexist jokes," being labelled "difficult" to work with while performing normal job functions or being questioned about their basic level of competence.*

team's opinions, micromanaging, fostering "old boys club" dynamics, or being completely aloof while the work falls on the team.

**Respondents reported a range of experiences with bias and/or discrimination in PIship and beyond. Some respondents felt that they experienced subtle or normalized forms of discrimination throughout their careers, while others felt like they experienced more punctuated and obvious acts against them.** For example, a woman who received career mentorship from a more senior woman said that she was given the advice that, to avoid gender discrimination throughout her career, she should be "the best in the room" in any given situation. The reasoning was that if she was more qualified and competent than the men around her, there would be lessened opportunities for discrimination to occur, including less doubt from team members as to her qualifications to serve as PI. The need to work harder than most others around her became a normalized part of her life.

**In another instance, a woman respondent felt that being passed over for a position was due to what she characterized as normalized gender bias at her place of work, despite her distinguished career accomplishments. Another woman perceived being the target of sexist jokes by colleagues while attempting to perform managerial tasks as a PI. Multiple women reported being labelled difficult to work with while performing normal job functions as PIs, and subsequently became targets of gossip as they gained reputations as being "hard to work with."**

A respondent identifying as a racial/ethnic minority reported feeling "invisible" at work as they were passed over for opportunities (such as serving on committees) for which they were uniquely qualified. In contrast, another respondent identifying as a racial/ethnic minority reported feeling heightened visibility at work when a white coworker reported them to a more senior colleague as having an abrasive personality and being non-collegial, formally recommending this person be denied a promotion.

While many respondents had suspicions of bias at points throughout their careers, several felt they experienced outright discrimination when they reached the level of PI. These more pointed experiences that were mentioned included receiving feedback on mission proposals that felt particularly personal, such as questioning if the PI could "handle" the job or doubting the PI's basic understanding of scientific concepts. In another instance, a respondent reported that while serving as PI she was frequently spoken over by older men on her mission team and had to resort to yelling on at least one occasion before being listened to. Women PIs reported that sometimes men refused to follow their directions for completing parts of the proposal and/or project, ignoring instructions and completing the tasks the way they saw fit.

Respondents had various reasons for not reporting such incidents. Some were hesitant to label certain incidents as intentional or deliberate acts of bias or discrimination. For example, one participant who identified as an ethnic/racial minority said they did not want to seem "ungrateful" for the opportunities they had been given by "complaining" about perceived racial bias in their career. Others reported that the pain, frustration, and difficulty of addressing these experiences are compounded by a perceived lack of ombud or protected reporting mechanism at NASA. One respondent's perception was that, because NASA carries such prestige, there is no governing body higher than the organization through which PIs could make complaints about discrimination or unfair treatment. Another respondent suggested the creation of a co-equal oversight body within NASA that had regulatory power to address grievances. Many respondents reported fearing retribution if they brought to light their experiences.

## CONCLUSION

NORC conducted research to understand the factors that impede or facilitate DEIA in the proposed leadership for competed space missions. The goal of this work was to support the Committee's efforts toward making recommendations to the NASA SMD about ways to overcome barriers to diversity, equity, inclusion, and accessibility, and ensure a diverse pool of proposed leadership. For this, NORC conducted semi-structured interviews with 29 individuals who had experience preparing and submitting proposals as PIs to the NASA SMD competed space missions program from 2010 to the present.

This study is an important first step in understanding barriers to diversity and inclusion in competed space missions funded by the NASA SMD. However, as indicated previously in the Limitations section, this study was unable to accommodate the full range of barriers individuals face, since it focuses only on the role of a PI. Several individuals who were not eligible for the study emailed the NORC team to express a concern that the study was too narrowly focused on PIs. They felt that this eligibility criterion excluded many others who face significant "barriers to entry" to proposal submission, and that it was equally important to hear from individuals who have been unable to make even a single submission. They were also concerned that a focus only on PIs would miss important factors shaping DEIA in competed space missions that impact individuals at other career stages. In addition to this feedback received from non-participants, several study participants, mostly comprised of women and minorities, also expressed concern about the limited scope of the study given their experiences in the field. Therefore, while feedback received by NORC from the broader space sciences community as well as from study participants showed enthusiastic support for this work, there is a need for additional studies on this topic such as the following: (1) those that seek to understand barriers to proposal submission faced by individuals who never reach the level of PI, (2) those that seek to understand barriers faced by others in leadership roles such as Deputy PI or Co-I, and (3) those that seek to understand barriers faced by proposers of instrument missions.

Despite these limitations, this study found a range of perceived barriers and facilitators that shape the proposed leadership for competed space missions. For example, respondents reported that higher education did not confer all the skills that they needed to become a successful PI, but that it served as a critical time to gain soft skills and mission experience and to build networks needed for career growth. Making connections and building networks with other scientists on missions was seen as critical to gaining opportunities later on. In addition, mentorship experiences were foundational to respondents' sense of belonging in the field, but they noted that this currently depended on the chance of meeting a mentor who was motivated to help the mentee. As such, respondents suggested that the culture of the scientific community needs to shift to encourage experienced PIs to mentor a wider swath of students and early career scientists.

Respondents also reported that preparing a proposal was extremely taxing and required strong personal, financial, and institutional supports. This included support from family members, support from their academic institutions (particularly for those who were pre-tenure), and financial support for their time while they prepared the proposal. Factors such as institutional affiliation and personal reputation impacted how teams were assembled as they shaped differential access to information, timelines, and processes. For example, respondents felt that those who work at NASA centers have a better understanding of NASA reporting requirements, budget monitoring, and other administrative processes. Others felt that industry partners are not always explicit in how partnerships form.

Finally, respondents explained that it is common for a proposal to be rejected during several rounds before being funded, due to the limited amount of funding for missions and a perception that mission concepts need to reach maturity. Given the vast financial scale of space missions, many characterized NASA as being more "conservative" with risk taking regarding PI selection, proposal budgets, and industry partners. Sometimes after a rejection respondents received thorough feedback from reviewers which they viewed as helpful in improving their proposal for the next round. Other times, respondents reported that the feedback was cursory and unhelpful, and they felt unsure if the reasons stated for rejection actually reflected behind-the-scenes decision-making. Because respondents spend much time and effort preparing and submitting a proposal, these efforts could result in career set-backs, financial risks, and personal sacrifices.

Intersecting these barriers and facilitators were other factors such as race/ethnicity and gender identity. Women and people who identified as belonging to ethnic/racial minorities reported less positive mentorship and early career experiences, cited increased experiences with subtle and overt acts of discrimination or bias throughout their careers, and shared the perception that proposal outcomes and experiences with PIship were negatively shaped by their gender, race, and/or ethnicity.

# D

# Committee and Staff Bios

## COMMITTEE

FRANCES (FRAN) BAGENAL, *Co-Chair*, is a senior research scientist and the assistant director for planetary sciences in the Laboratory for Atmospheric and Space Physics (LASP) at the University of Colorado Boulder, leading its Magnetospheres of the Outer Planets Group. She is also a professor emerita in the Department of Astrophysical and Planetary Sciences at the University of Colorado Boulder. Her research interests focus primarily on the outer solar system, explicitly understanding planetary magnetospheres by combining data analysis and theoretical models. She was a co-investigator (Co-I) on the Plasma Science instrument on the Voyager mission; an interdisciplinary scientist on the Galileo mission; a Deep Space 1 team member; a New Horizons mission Co-I and Particles Theme Lead; and a Juno Co-I and co-chair of the Magnetospheres Working Group and the Science Planning Working Group. She has been active in workforce issues, including serving on the advisory committees for the 2011 and 2020 Planetary Science Workforce Surveys. She is the recipient of numerous awards, including the James Van Allen Lecture Award from the American Geophysical Union and the Boulder Faculty Assembly's Excellence in Research Award. She is a member of the National Academy of Sciences (NAS) and a fellow of the American Geophysical Union (AGU) and the American Astronomical Society (AAS). She earned her Ph.D. in Earth and planetary sciences from the Massachusetts Institute of Technology (MIT).

WANDA E. WARD, *Co-Chair*, is the executive associate chancellor for public engagement at the University of Illinois at Urbana-Champaign (UIUC), where she provides strategic leadership for the university and the broader community to advance the university's land-grant, public engagement mission. Additionally, Ward chairs the COVID-19 Return to On-Campus Operations: Community and Public Engagement Committee; co-chairs the Call to Action to Address Racism and Social Injustice: Community Action and Public Engagement Committee; and is a member of the Carle Foundation (Urbana) and the Champaign County Community Coalition Executive Committee (Champaign). She joined the university in 2018 as the executive associate chancellor for administration and university relations, where she served primarily as the Chief of Staff (2018-2021). Before joining UIUC, Ward served in several senior leadership positions throughout her 26 years at the National Science Foundation (NSF), including as the senior advisor to the director; the head of the Office of International and Integrative Activities; the deputy assistant director for social, behavioral, and economic sciences; and the deputy assistant director for education and human resources. Also, she has served as the assistant director for broadening participation at the Office of Science and Technology Policy (OSTP), the White House Executive Office of the President; on the

U.S. President's National Science and Technology Council subcommittees and interagency working groups in the areas of the social, behavioral, and economic sciences; science education; and workforce development; and at the international level, as a member of the International Social Science Council (ISSC) Committee for Developing and Transition Economies (CoDATE) and the ISSC Executive Committee. She was awarded the American Psychological Association (APA) Presidential Citation, the Federation of Behavioral, Psychological, and Cognitive Sciences Richard T. Louttit Award, and the U.S. Presidential Rank Award of Distinguished Executive. Ward was also selected as a fellow in the University of Illinois System President's Executive Leadership Program. She earned a B.A. in psychology with an Afro-American Studies Certificate from Princeton University and a Ph.D. in psychology from Stanford University.

CLAUDE R. CANIZARES is the Bruno Rossi Professor of Physics at MIT, where he has also served as the vice president, the vice president for research, the associate provost, and the director of the Center for Space Research. Canizares is the associate director of the Chandra X-ray Observatory Center and a principal investigator (PI) on NASA's Chandra X-ray Observatory, having led the development of the Chandra High Resolution Transmission Grating Spectrometer. Canizares's main research interests are high-resolution x-ray spectroscopy and plasma diagnostics of supernova remnants and clusters of galaxies; X-ray studies of dark matter; X-ray properties of quasars and active galactic nuclei; and observational cosmology. He is the author or co-author of more than 230 scientific papers. Canizares is a member of the NAS and has received several awards, including decoration for Meritorious Civilian Service to the United States Air Force, two NASA Public Service Medals, and the Goddard Medal of the AAS. Canizares earned a B.A., M.A., and Ph.D. in physics, all from Harvard University.

ERIN CECH is an assistant professor of sociology at the University of Michigan. Previously, she was a post-doctoral fellow at the Clayman Institute for Gender Research at Stanford University and was on faculty at Rice University. Cech's research examines cultural mechanisms of inequality reproduction, especially through seemingly innocuous cultural beliefs and practices. Her work on inequality in science, technology, engineering, and mathematics (STEM) professions focuses on the recruitment and retention of women, people of color, and LGBTQ-identifying persons in STEM degree programs and STEM jobs, and the role of professional cultures in this inequality. Cech examines how cultural definitions of "good work" and "good workers" can anchor inequality in the workforce. For example, she examines the role of the "passion principle" in the reproduction of occupational inequalities: how seemingly voluntary and self-expressive career decisions help reproduce processes like occupational sex segregation. She also studies how cultural understandings of the extent and origin of inequality help to uphold unequal social structures. She has worked with NASA previously (Goddard and Johnson Space Center) on diversity issues as it relates to LGBTQ+ populations. Cech's work is funded by multiple grants from NSF, and her research has appeared in the *Proceedings of the National Academy of Sciences*, the *American Journal of Sociology*, and the *American Sociological Review*. Her research has also been covered by *The New York Times*, *The Washington Post*, CNN, *Time*, *Harvard Business Review*, and the news sections of *Science* and *Nature*. In 2020, she was named one of *Business Equality* Magazine's "40 LGBTQ+ Leaders Under 40" and was honored with the University of Michigan's Henry Russel Award. She earned a Ph.D. in sociology from the University of California, San Diego.

ANNMARIE ELDERING is the deputy project scientist for the Orbiting Carbon Observatory-2 (OCO-2) and the project scientist for OCO-3 at the Jet Propulsion Laboratory (JPL). She has been at JPL since 1999, adapting her knowledge of radiative transfer and light scattering to algorithm development for the Tropospheric Emission Spectrometer (TES) as well as validation of AIRS data. She served as the TES deputy PI and then the PI between 2006 and 2010. That led to leadership roles as the deputy section manager and section manager in Earth Atmospheric Sciences section before she joined the OCO-2 project in 2010. In her recent project and management roles, she provides dynamic technical leadership for teams focused on the extraction of information about clouds, aerosols, and trace gases in the Earth's atmosphere. She is the PI or a Co-I on a number of competitively awarded research tasks and has worked on the development of several mission concepts. She earned a Ph.D. in environmental engineering science from the California Institute of Technology (Caltech).

ELENA A. EROSHEVA is a professor of statistics and social work and the associate director of the Center for Statistics and the Social Sciences at the University of Washington. Her research focuses on the development and application of statistical methods and models for complex and heterogeneous data in the social, behavioral, medical, and health sciences. Her work includes statistical methodology development for multivariate and longitudinal data analysis, text analysis, survey methodology, and analysis of survey data. She is a first prizewinner of the 2014 America Competes Act Challenge competition to maximize fairness in National Institutes of Health (NIH) peer review in the category Most Creative Idea for Detection of Bias in Peer Review. In line with this award, she recently published findings on disparities in peer review scores between white and black applicants for NIH R01 grants. Erosheva is currently serving as an ArXiv moderator for *Statistics*, and her current and past editorial board service includes the *Annals of Applied Statistics*, the *Journal of the American Statistical Association*, the *Journal of Educational and Behavioral Statistics*, and *Psychometrika*. She earned a Ph.D. in statistics from Carnegie Mellon University.

ORLANDO FIGUEROA is the president at Orlando Leadership, LLC. He is retired from NASA where he served as the former NASA director for Mars Exploration, the director for the Solar System Division in the Office of Space Science at NASA Headquarters, and as the deputy center director for science and technology of the Goddard Space Flight Center. Figueroa's career with NASA spanned 33 years; throughout which he demonstrated a record of consistent achievements in the planning, design, development, and oversight of multi-faceted scientific space missions. He is well versed in interacting with national and international government and non-governmental organizations. He is considered a highly competent strategist and director of scientific space systems and technology, the management of large multi-disciplinary engineering organizations; and experienced in the management of complex multi-project programs with total budgets in excess of $4 billion. He earned a B.S. in mechanical engineering from the University of Puerto Rico, Mayaguez.

ALLISON N. JAYNES is an assistant professor of physics and astronomy at the University of Iowa, where she researches space and plasma physics. Her primary research interests include pulsating aurora, radiation belts, and the connection between the two. She is the Co-I on NASA's Magnetic Multiscale and Van Allen Probes missions, as well as on a NASA-funded CubeSat, GTOSat, and a NASA-funded sounding rocket mission, LAMP. She served on the Nomination Task Force within AGU's Space Physics and Aeronomy (SPA) section assembled to increase the diversity of honors and awards winners at AGU. She earned a Ph.D. in physics from the University of New Hampshire.

LINDSEY MALCOM-PIQUEUX is the assistant vice president for diversity, equity, inclusion, and assessment at Caltech. Malcom-Piqueux's primary interest centers on post-secondary access and success for minority women and men in STEM fields. Following this interest, she has conducted research in several areas including the relationship between financial aid and STEM outcomes, the role of 4-year minority-serving institutions and community colleges in creating educational equity in STEM, and gender equity in STEM among minoritized populations. Malcom-Piqueux also studies the educational experiences and outcomes of students of color at Historically Black Colleges and Universities (HBCUs), predominantly Black institutions, Hispanic-Serving Institutions, and Tribal Colleges and Universities; and examines the congruence of the minority-serving designation and academic outcomes among their target populations. Her work has appeared in *Educational Researcher*, the *Review of Higher Education*, and the *Harvard Educational Review*, among other journals, and in volumes edited by Routledge, SUNY Press, Johns Hopkins University Press, and Stylus Publishing. She recently authored a commissioned paper for the National Academies Symposium on Imagining the Future of Undergraduate STEM Education titled "Transformation in the U.S. Higher Education System: Implications for Racial Equity." Additionally, she has previously served on the National Academies study on Developing Indicators for Undergraduate STEM Education. Malcom-Piqueux earned a Ph.D. in urban education with an emphasis on higher education from the University of Southern California, and an M.S. and an S.B. in planetary science from Caltech and MIT, respectively.

DAVID J. MCCOMAS is the vice president for the Princeton Plasma Physics Laboratory and a professor of astrophysical sciences at Princeton University. He has participated in numerous solar science missions, including as the PI for NASA's IMAP, IBEX, and TWINS missions and a lead Co-I for instruments on the ACE and New Horizons missions. McComas has invented a variety of instruments and missions for space applications and has authored papers in the fields of heliospheric, magnetospheric, solar, and planetary science, as well as on space instrument and mission development. His numerous honors include the 2014 COSPAR Space Science Award, a NASA Exceptional Public Service Medal, and AGU's James B. Macelwane Medal. McComas serves on the board of directors of the Brookhaven National Laboratory Science Associates. He recently served on the NASA Advisory Council (NAC) and chaired the NAC Science Committee. Previous service includes chairing NASA's Sun-Earth Connections Advisory Subcommittee, chairing two Solar Probe and Solar Probe Plus Science and Technology Definition Teams, and serving as a member of NASA's Space Science Advisory Committee. He earned a B.S. in physics from MIT and a Ph.D. in geophysics and space physics from the University of California, Los Angeles.

CHRISTOPHER B. NEWMAN is an associate professor in the Department of Higher Education at Azusa Pacific University (APU). Before joining the faculty at APU he was a professor at the University of San Diego. His research focuses primarily on outcomes, inequities, and undergraduate student experiences in STEM. Additionally, he has interests in multi-cultural education in global contexts. Utilizing critical leadership and organizational change theoretical perspectives, Newman has worked with colleges and universities in the United States and abroad as they prepare for diversity, inclusion, equity, and/or racial justice action. He has worked, in collaboration with the National Society of Black Physicists (NSBP), on African American workforce development in physics and astronomy. Newman has served as a consultant to NSF's Colloquy on Minority Males in STEM and regularly serves as a review panelist for grant proposals to NSF. Newman earned a B.S. in sociology from the University of California, Santa Barbara, an M.S. in leadership studies from the University of San Diego, and a second M.S. and a Ph.D. in higher education and organizational change from the University of California, Los Angeles.

KUNIO M. SAYANAGI is an associate professor in the Department of Atmospheric and Planetary Sciences at Hampton University. His core research expertise is in atmospheric dynamics. He applies his expertise to understand the atmospheres of solar system planets through theoretical analyses; numerical modeling; Earth-based telescopic observation analyses, space-based remote-sensing, and in-situ measurements; and planetary mission concept development. He was the PI of the Small Next-generation Atmospheric Probe mission concept development funded by NASA's PSDS3 program. He was also an affiliate member of the NASA Cassini mission's Imaging Science Team. He has also been active in formulating and advocating for NASA's planetary exploration priorities and is currently on the Steering Committee of NASA's Outer Planets Assessment Group. He was selected to be on the first Forum for New Leaders in Space Science program in 2014 jointly organized by the Chinese Academy of Sciences and the Space Studies Board of the National Academies. He leverages his position as a professor at an HBCU to broaden participation in space science, and actively recruits undergraduate students to start research careers in space science while they are in college. He also regularly leads and participates in regional educational/public outreach activities at local schools and science museums. He earned a Ph.D. in physics from the University of Arizona.

KEIVAN G. STASSUN is the Stevenson Professor of Physics and Astronomy at Vanderbilt University. Stassun served on the AAS task force for promoting diversity and inclusion in astronomy graduate education. He is known for his leadership and distinction as a scientist and as an innovator in broadening the participation of underrepresented minorities in STEM fields. He is the founding director of the Fisk-Vanderbilt Master's to Ph.D. Bridge Program, is a recipient of the Presidential *Awards* for Excellence in Science, Mathematics, and Engineering *Mentoring* award, and has given testimony before the U.S. House of Representatives on the Bridge program and broadening participation in STEM fields. He has extensive National Academies committee experience, and specifically on projects requested by NASA, including the Decadal Survey on Astronomy and Astrophysics 2020 (Astro2020), the Panel on the State of the Profession and Societal Impacts in Astronomy, the Committee on the NASA Science Mission Directorate Science Plan, the Committee on Effective Mentoring in STEMM, and the

Committee on Understanding and Addressing the Underrepresentation of Women in Particular Science and Engineering Disciplines. He earned a Ph.D. in astronomy from the University of Wisconsin–Madison.

LOUIS-GREGORY STROLGER is the deputy head of the instruments division and observatory scientist in the Science Mission Office at the Space Telescope Science Institute. He is primarily concerned with clues to the nature of supernova progenitors through bulk analyses, rates, environmental effects (star-formation, metallicity, etc.), and the global evolution of these properties over the history of the Universe. He is also involved in projects on four robotic telescopes: three in space, the Hubble Space Telescope, the James Webb Space Telescope, and the Nancy Grace Roman Space Telescope; and one on the ground, the RCT 1.3-meter. Strolger is active in several initiatives addressing underrepresented minorities in astronomy and physics, and looking at approaches to improve recruitment and retention. He earned a Ph.D. in astronomy and astrophysics from the University of Michigan, Ann Arbor.

## STAFF

SANDRA J. GRAHAM (*Co-Study Director*) has served as a senior program officer at the National Academies of Sciences, Engineering, and Medicine's Space Studies Board (SSB) since 1994. During that time Graham has directed a large number of major studies, many of them focused on space research in biological and physical sciences and technology, including the comprehensive 2011 decadal survey, Recapturing a Future for Space Exploration—Life and Microgravity Sciences Research for a New Era. High-profile studies in other areas have included an assessment of servicing options for the Hubble Space Telescope, a study of the societal impacts of severe space weather, and the Pathways to Exploration report reviewing the U.S. human space program. Prior to joining the SSB, Graham held the position of senior scientist at the Bionetics Corporation, where she provided technical and science management support for NASA's Microgravity Science and Applications Division. Graham's honors include the National Research Council's (NRC's) Distinguished Service Award, the NRC Division on Engineering and Physical Sciences Exceptional Achievement Award, and the Orr-Reynolds Distinguished Service Award from the American Society for Gravitational and Space Research. She is a member of the American Chemical Society and the Project Management Institute. She received her Ph.D. in inorganic chemistry from Duke University, where her research focused on rate modeling and reaction chemistry for biological metal complexes and their analogs.

TIFFANY E. TAYLOR (*Co-Study Director*) is a program officer for the Board on Science Education (BOSE) at the National Academies of Sciences, Engineering, and Medicine. In this role, she provides research, planning, and management support for several ongoing projects including the Standing Committee on Advancing Science Communication, a consensus study on Enhancing Science and Engineering in Grades PreK-5, and a consensus study on Equity in PreK-12 STEM Education. Additionally, Taylor is the co-study director for a consensus study requested by NASA on Increasing Diversity and Inclusion in the Leadership of Competed Space Missions, in collaboration with the SSB. Taylor came to the National Academies as a Christine Mirzayan Science and Technology Policy fellow in 2017, where she also worked with the BOSE. She received her bachelor's degree in biology from Howard University and her Ph.D. in biomedical sciences from the University of California, San Diego. Taylor is extremely passionate about the inclusion of persons of diverse background in science, and aspires to leverage her Ph.D. training and science policy experience to address educational inequity within society, in both domestic and global settings.

ABIGAIL SHEFFER is a senior program officer with the SSB. She has been the staff officer and study director on a variety of activities such as the Decadal Survey on Astronomy and Astrophysics (Astro2020), the Committee on Solar and Space Physics, Open Source Software Policy Options for NASA Earth and Space Sciences, Strategic Investments in Instrumentation and Facilities for Extraterrestrial Sample Curation and Analysis, and Achieving Science with CubeSats: Thinking Inside the Box, among others. In 2009, Sheffer came to the National Academies as a Christine Mirzayan Science and Technology Policy Graduate Fellow with the SSB. Sheffer earned her Ph.D. in planetary science from the University of Arizona and A.B. in geosciences from Princeton University.

GAYBRIELLE HOLBERT joined the SSB and the Aeronautics and Space Engineering Board (ASEB) as a program assistant in 2019. In this role, she assists with ongoing projects and workshops by providing logistics, report creation, and project support including the Standing Committee on Earth Sciences and Applications from Space (CESAS). She previously worked as a communication specialist for the United Black Fund Inc. a non-profit organization that helped inner-city youth by providing after-school programs, scholarships, and resources to engage and enhance their educational needs. Prior to that, she was the social media consultant for the Development Corporation of Columbia Heights, a non-profit advocacy platform that supports responsible, community-focused economic development and as a production assistant for a startup multimedia production company. She holds a B.A. in mass media communications from the University of the District of Columbia.

COLLEEN HARTMAN joined the National Academies in 2018, as the director for both the SSB and the ASEB. In 2021, she additionally became the director for the Board on Physics and Astronomy. After beginning her government career as a presidential management intern under Ronald Reagan, Hartman worked on Capitol Hill for House Science and Technology Committee Chairman Don Fuqua as a senior engineer building spacecraft at NASA Goddard and as a senior policy analyst at the White House. She has served as Planetary Division Director, Deputy Associate Administrator and Acting Associate Administrator at NASA's Science Mission Directorate, as Deputy Assistant Administrator at the National Oceanic and Atmospheric Administration, and as Deputy Center Director and Director of Science and Exploration at NASA's Goddard Space Flight Center. Dr. Hartman has built and launched scientific balloon payloads, overseen the development of hardware for a variety of Earth-observing spacecraft, and served as NASA program manager for dozens of missions, the most successful of which was the Cosmic Background Explorer (COBE). Data from the COBE spacecraft gained two NASA-sponsored scientists the Nobel Prize in physics in 2006. She also played a pivotal role in developing innovative approaches to powering space probes destined for the solar system's farthest reaches. While at NASA Headquarters, she spearheaded the selection process for the New Horizons probe to Pluto. She helped gain administration and congressional approval for an entirely new class of funded missions that are competitively selected, called "New Frontiers," to explore the planets, asteroids, and comets in the Solar System. She has several master's degrees and a Ph.D. in physics. Hartman has received numerous awards, including two prestigious Presidential Rank Awards.

HEIDI SCHWEINGRUBER is the director of the BOSE at the National Academies. She has served as the study director or the co-study director for a wide range of studies, including those on revising national standards for K-12 science education, learning and teaching science in grades K-8, and mathematics learning in early childhood. She also co-authored two award-winning books for practitioners that translate findings of National Academies' reports for a broader audience, on using research in K-8 science classrooms and on information science education. Prior to joining the National Academies, she worked as a senior research associate at the Institute of Education Sciences in the Department of Education. She also previously served on the faculty of Rice University and as the director of research for the Rice University School Mathematics Project, an outreach program in K-12 mathematics education. She has a Ph.D. in psychology (developmental) and anthropology and a certificate in culture and cognition, both from the University of Michigan.